Workbook for

The Human Body in Health and Disease

Workbook for

The *Human Body* in *Health* and *Disease*

Second Edition

Ruth Lundeen Memmler, M.D.

Professor Emeritus, Life Sciences;
formerly Coordinator, Health, Life Sciences and Nursing,
East Los Angeles College, Los Angeles

and

Dena Lin Wood, R.N., B.S., P.H.N.

Assistant Head Nurse, Los Angeles County—
University of Southern California
Medical Center, Los Angeles

Illustrated by Anthony Ravielli

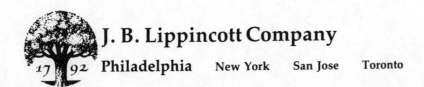

J. B. Lippincott Company

1792 **Philadelphia**   New York   San Jose   Toronto

Preface

This workbook has been revised so that it may be used most effectively with the fourth edition of *The Human Body in Health and Disease*. As in the first edition the chief objective is to aid the beginning student in learning the fundamentals of body structure and function as well as some basic medical terminology. It is expected that this workbook may also be used in connection with other textbooks that include biological and physical principles related to the human organism.

If the student will practice pronouncing the scientific words while writing them in answer to the questions in the workbook he should find that his understanding of the material is increased. Studying the questions and their answers should help the student understand structure and function of the human body, as well as various common disorders. Comparing the normal with the abnormal usually helps fix in the student's mind the basic facts that underlie care and cure processes. Obviously he gains the most by doing his own work while avoiding the copying of a fellow student's material.

The questions are written with the aim of assisting the student in the learning process. They are not intended as a test of his knowledge. Sometimes the answer to one question is purposely included in another interrogation in order to reinforce the ideas that are presented. The student should analyze the facts and try to place them in a meaningful order.

Anthony Ravielli illustrated the book. The authors are grateful for suggestions and help given by colleagues and by the staff of the J. B. Lippincott Company, especially Bernice Heller and David T. Miller.

Contents

Introduction to Medical Terminology

I. OVERVIEW

Medical terminology is the special language of the health occupations. This terminology is concerned with the human body. It deals with the parts of the body (anatomy) and their functions (physiology), as well as with the normal and abnormal states of the body, the disorders and injuries to which it is subject, and the various means employed to maintain the body's normal state or to correct an abnormal state as far as possible.

Medical terminology is a necessary tool for the student who is interested in medical science and who hopes to be employed in one of the numerous fields of endeavor that are related to it. This terminology is a kind of shorthand; just as the single word *book* describes a set of printed or written sheets bound together into a volume, so the single word *meningitis* (men-in-ji'tis) describes a condition in which there is inflammation of meninges, or brain coverings. This terminology should be instantly understood by everyone who is involved in any fashion in one of the health occupations.

The vocabulary of medical science is a large one, since it reflects all that is known and identified about the human body. Yet the similarities that exist among many of these words help make it possible for the student to understand and use the language of medical science, and to feel at ease with this language.

Most medical words are made up of 2 or more components, or parts. The *root*, or *combining form*, is, like the root of a tree, the anchor to which the other part or parts are attached. These other parts are designated *prefixes* (coming before the root), or *suffixes* (coming after the root), and adjective or noun endings (indicating a relationship to the root). A *compound word* is, as the term suggests, made up of 2 or more roots plus one or more of the other parts just mentioned. To understand the meaning of each word, first divide the word into its parts and then look up the meaning of each part. When these parts are put together, the complete word becomes understandable. You will find it helpful to pronounce each syllable individually first, and then say the whole word several times, while concentrating on its meaning. Examples:

1

1. hyperthermia (hi-per-ther′me-ah) = fever, a condition of increased heat in the body
    a. *prefix* (hyper = increased or above normal)
    b. *root* (therm = heat)
    c. *suffix* (ia = condition or state of being)
2. abdominohysterectomy (ab-dom-i-no-his-ter-ek′to-me) = removal of the uterus through the abdominal wall
    a. *combining form* (abdomino = the belly or abdomen)
    b. *combining form* (hystero = uterus or womb)
    c. *suffix* (ectomy = removal of)

II. TOPICS FOR REVIEW

1. common word roots and combining forms such as:

abdomin-, abdomino- *belly*	gastr-, gastro- *stomach*
aden-, adeno- *glands*	gynec-, gyneco- *women*
arthr-, arthro- *joint*	hem-, hema-, hemato-, hemo- *blood*
bio- *study of life*	hist-, histio- *tissue*
carcin-, carcino-	hyster-, hystero- *uterus*
cardi-, cardio- *heart*	idio- *self*
cephal-, cephalo-	lact-, lacto- *milk*
chole- *gall bladder*	leuc- or leuk-, leuko- *white blood cells*
chondr-, chondro- *cartilege*	neph-, nephro- *kidney*
cleid-, cleido-	neuro- *nerves*
cost- *ribs*	psych-, psycho- *mind*
cyt-, cyto- *cell*	somat-, somato- *body*
derm-, derma- *skin*	vas-, vaso- *vessel*

2. common prefixes (at the beginnings of words) such as:

a-, an-	mal-
ab-	meg-, mega-, megalo-
circum-	met-, meta-
contra-	micro-
di-	neo-
ex-	semi-
infra-	sub-
inter-	trans-
intra-	tri-
macro-	uni-

3. common suffixes (word endings) such as:

-algia	-itis
-cele	-logy, -ology
-ectasis	-oma
-ectomy	-ostomy
-esthesia	-otomy
-ferent	-penia
-gen	-phagia, -phagy
-geny	-plasty
-gram	-ptosis
-graph	-pnea

4. common adjective endings such as -ous and -al
5. common noun endings including -us and -um

2

III. MATCHING EXERCISES

Matching only within each group, print the answer in the space provided.

Group A

prefix	suffix	root
-ous, -al	-cele	a-, an-
compound word	combining form	

1. The foundation of a word is its .... _root_.

2. When 2 or more word foundations are used the result is a _compound word_

3. The part of a word that precedes its foundations and changes its meaning is a .... _prefix_ / ~~combining form~~

4. A word ending used to change the meaning of the word foundation is a .... _suffix_.

5. Examples of endings that indicate the adjective forms are _-ous, -al_.

6. The word root followed by a vowel (to make pronunciation easier) is a .... _combining form_ / ~~prefix~~

7. A suffix that means a swelling or an enlarged space is .... _cele_.

8. To denote absence or deficiency begin the word with prefixes such as .... _a-, an-_.

Group B

psych-	abdomin-	-algia
cyt-	hema-	somat-
hist-	aden-	neo-

1. To indicate the belly area use .... _abdomin-_

2. A word root that means gland is .... _aden-_

3. To show relationship to a cell use .... _cyt-_

4. A word root for tissue is .... _hist-_

5. Relationship to mind is shown by .... _psych-_

6. A word part that means blood is .... _hema-_

7. A word root that indicates body is .... _somat-_

8. A prefix that means new is .... _neo-_

9. A suffix that refers to pain is .... _-algia_

3

Group C

arthr-	-itis	carcin-
-esthesia	infra-	-ptosis
meg-	-ectasis	-ectomy

1. A prefix that indicates excessively large is .... _meg-_ .

2. A suffix that means downward displacement is .... _-ptosis_ .

3. To indicate inflammation, end the word with .... _-itis_ .

4. To show relationship to a joint use .... _arthr-_ .

5. Dilation or expansion of a part is indicated by the word ending .... _-ectasis_ .

6. Removal or destruction of a part is shown by the addition of .... _-ectomy_ .

7. To refer to sensation use the suffix .... _-esthesia_ .

8. To indicate a cancer use .... _carcin-_ .

9. To show that a part is located below use the prefix ...... _infra-_ .

Group D

-us, -um	-ous, -al	-otomy - _opening_
leuko-	-genic	-penia
-ostomy	erythr-	ab- _away_

1. To indicate producing add .... _-genic_ .

2. A lack of is shown by the suffix .... _-penia_ .

3. A suffix that means incision into a part is .... _-otomy_ .

4. To indicate the formation of a new opening use .... _-ostomy_ .

5. To show that something is red use the word part .... _erythr-_ .

6. To indicate that something is white use .... _leuko-_ .

7. A prefix that means away from is .... _ab-_ .

8. Endings that show the adjective form are .... _-ous, -al_ .

9. Noun forms of words may end in .... _-us, -um_ .

4

Group E

Combine appropriate word parts from the list below and print the correct word in the blanks.

hemo- or hemato-	aden- *gland*	-logy
-costal	arthr-	oste- or osteo- *bones*
inter- *inside*	-lysis	carcin- *cancer*
-ectomy -*removal*	bio- *living*	-cellular
-itis	-oma	intra-
cyto-		

1. The study of living things is called .... _biology_ .

2. Inflammation of a joint is known as .... _arthritis_ .

3. Removal of a gland is called .... _adenectomy_.

4. The scientific study of cells is known as .... _cytology_ .

5. The removal of a joint is called .... _arthrectomy_.

6. The space between the ribs is .... _intercostal_ .

7. A cancerous tumor is called a .... _carcinoma_.

8. The study of blood and its constituents is .... _hematology_.

9. A tumor filled with blood is a .... _hematoma_.

10. The word that means between cells is .... _intercellular_ .

11. The word that means inside of or within a cell is .... _intracellular_

12. A tumor made of glandular kinds of tissue is an .... _adenoma_ .

13. The dissolution or disintegration of blood cells (especially red blood cells) is called .... _hemolysis_ .

14. Destruction or dissolution of body cells may be called ... _cytolysis_ .

15. A firm tumor made of bone or bonelike tissue is known as an .... _osteoma_ .

Group F

Combine 2 or 3 word parts selected from the list below in order to correctly complete the following sentences. Print the appropriate words in the spaces provided.

electro-	micro-	trans-
cardio-	encephalo-	an-
-graph	-gram	-cyte
-emia	-orbital	-esthesia

5

1. A lack of or decrease in red blood cell constituents is called .... *anemia*.

2. An abnormally small red blood cell is a(n) .... *microcyte*.

3. A procedure performed through the bony eye socket is described by the word .... *trans orbital*.

4. An x-ray film of the head (including the brain) is called an *encephalgram*

5. The instrument that is used for producing a graphic tracing of heart muscle electric current is an .... _____.

6. A graphic record of electrical currents in the brain (brain waves) is called an .... *electroencephalograph*

7. The instrument used for producing a graphic record of brain waves is the .... _____.

8. The tracing of heart muscle electric current is called an .. *electrocardiogram*

9. A lack of or loss of sensation (especially of pain) is designated .... *anestesia*.

IV. COMPLETION EXERCISE

Print the word or phrase that correctly completes the sentence.

1. A prefix that indicates very small size is .... *micro*.

2. Words that refer to something written or recorded end with .... ~~graph~~ *graph*.

3. The visible record produced by recording electrical currents is indicated by a word ending in .... *gram*.

4. A prefix that denotes below or under is .... *sub -/hypo-*

5. To show that something is outside or is sent outside use the prefix .... *ex*.

6. To indicate that there are 3 parts to an organ begin the word with the prefix .... *tri*.

7. A prefix that means across, through or beyond is .... *trans*.

8. To indicate surgical molding, use the suffix .... *plasty*.

9. The prefix that shows something is within the structure is *intra*.

10. The noun form of the adjective mucous is .... _____.

6

11. Suffixes that indicate the process of eating or swallowing include .... *phajia*.

12. The suffix that means bladder or sac is .... *cele*.

13. The prefix that means away from is .... *ab*.

14. The suffix that means painful is .... *algia*.

15. A suffix that means removal by surgery or other means is _____.

16. A 2-letter prefix that means absence or lack of is .... *an*.

17. A suffix that means dilation or expansion of a part is .... _____.

18. An agent that produces or originates is indicated by the suffix .... *gen*.

19. The word root for tissue is .... *hist*.

20. Prefixes that mean excessively large include .... *megalo*.

V. PRACTICAL APPLICATIONS

Study each discussion. Then print the appropriate word or phrase in the space provided.

1. Mr. A, age 74, was admitted because of weakness and inability to care for himself. The physician noted that Mr. A suffered from malnutrition. Later studies showed there was evidence of malunion of a fracture of the right thigh bone. The prefix mal means .... *lack of*.

2. Baby John was brought to the clinic by his observant mother because one of his eyes did not seem normal. The doctor noted that there was unilateral enlargement of the right pupil and that this uniocular condition would require laboratory investigation. The prefix uni- means .... _____.

3. Mrs. B was admitted for treatment of an injured hand. The admitting intern noted that examination of the metacarpal bones showed possible fractures. The prefix meta- means .... _____.

4. Miss C was examined in the outpatient department. It was noted that she had circumoral pallor and that this pallor was circumscribed. The prefix circum- means ..... _____.

5. Later examination of Miss C showed ptosis (to'sis) of several abdominal organs, a condition called visceroptosis (vis-er-op-to'sis). The word part ptosis may also be used as a separate word. It means .... _____.

VI. REFERENCES

Memmler, R. L., and Wood, D. L.: The Human Body in Health and Disease, ed. 4, pp. 291-295. Philadelphia, Lippincott, 1977.

Chaffee, E. E., and Greisheimer, E. M.: Basic Physiology and Anatomy, ed. 3, pp. 527-530. Philadelphia, Lippincott, 1974.

Frenay, A. C.: Understanding Medical Terminology, ed. 5. St. Louis, Catholic Hospital Association, 1973.

Greisheimer, E. M., and Wiedeman, M. P.: Physiology and Anatomy, ed. 9, pp. 634-646. Philadelphia, Lippincott, 1972.

Patrick, F. J.: Basic Medical Terminology, Text, Manual and Tape Cassette. New York, Bobbs Merrill, 1975.

CHAPTER ONE

The General Plan of the Human Body

I. OVERVIEW

In order to understand the general organization of the human body we may begin with the smallest "bricks," or *cells*, which are the basic structural units of all living things, and which are composed of protoplasm. Although there are numerous types of cells, which differ in size and attributes according to function, all have a common basic structure. The outer *cell membrane* contains the main substance, or *cytoplasm*; in the center of the cell is the *nucleus*, which contains the still smaller *nucleolus*. Cells combine to form tissues that in turn form *organs*; these organs form *systems*.

It is essential that a special set of terms be learned in order to locate parts and to relate the various parts to each other. Imaginary lines called *planes of division* separate parts of the body into *regions* in much the same way that the equator, the Tropics of Cancer and Capricorn and the Arctic and Antarctic Circles divide the earth into zones. Further divisions of the earth by lines of latitude and longitude make it possible to pinpoint locations accurately. Similarly, separation into areas and regions within the body, together with the use of the special terminology for directions and locations, makes it possible to describe an area within the human body with considerable accuracy.

The logical, decimal metric system is now replacing all other systems of measurement. If you learn to "think metric," you will find it easier to use than the older systems in use in the United States.

II. TOPICS FOR REVIEW

1. microscopic structure
   a. cells and protoplasm
   b. nucleus and other cell parts
2. body planes, cavities and systems
3. body directions
   a. anatomical position
   b. superior and inferior
   c. ventral and dorsal
   d. anterior and posterior

e. cranial and caudal
f. medial and lateral
g. distal and proximal

III. MATCHING EXERCISES

Matching only within each group, print the answer in the space provided.

Group A

organs   tissues   cells
systems   cytoplasm   nucleus
protoplasm

1. The substance of which all living things are composed is _protoplasm_.

2. The central, usually oval, part of a cell is the .... _nucleus_.

3. A combination of specialized groups of cells forms ...... _tissues_.

4. The main substance of the cell outside the nucleus is the _cytoplasm_.

5. A combination of various tissues forms parts having a special function called .... _organs_.

6. Several different parts and organs grouped together for specific functions form .... _systems_.

7. The building blocks of which living organisms are made are called .... _cells_.

Group B

epigastrium   ventral   distal
umbilicus   lateral   medial
proximal   thoracic region   transverse

1. To indicate nearness to the midsagittal plane use the word .... _medial_.

2. A part that is away from the midline (or toward the side) is .... _lateral_.

3. To indicate that a part is <u>near</u> or toward the point of origin use .... _distal proximal_.

4. A part that is <u>away</u> from the point of origin is .... _proximal distal_.

5. A horizontal or cross section is also said to be .... _transverse_.

6. The central region of the abdomen just below the breast bone is the .... _epigastrium_.

7. Another name for the navel is the .... _umbilicus_.

10

8. The upper or chest portion of the ventral body cavities is the .... _thoracic region_

9. The word that means toward the belly surface is .... _ventral_ .

Group C

| caudal | cranial | posterior |
| ventral | proximal | lateral |

1. To say, toward the origin of a part, use the word .... _proximal_ .

2. To indicate that a part is toward the rear use .... _posterior_

3. The word that means nearer the tail region is .... _caudal_ .

4. To indicate that a part is nearer the head use the word .. _cranial_ .

5. To show that a part is toward the side use the word ..... _lateral_ .

6. To show that a part is nearer the belly area use .... _ventral_ .

Group D

| urinary | integumentary | - skeletal |
| endocrine | reproductive | respiratory |

1. The system that includes the hair, nails and skin is the ... _integumentary_

2. The bones, joints and related parts form the system called the .... _skeletal_ .

3. Another name for the excretory system is the .... _urinary_ .

4. The system of scattered organs that produce hormones is called the .... _endocrine_ .

5. The system that includes the sex organs is the .... _reproductive_

6. The lungs and bronchial tubes form the system called the _respiratory_

Group E

| spinal canal | cranial cavity | diaphragm |
| midsagittal | chromatin network | metric system |

1. A decimal system of measurement is the .... _metric system_

2. The plane that divides the body into right and left halves is the .... _midsagittal_

3. The lower elongated part of the dorsal body cavity is the _spinal cavity_

4. The muscular partition between the 2 ventral body cavities is the .... _diaphram_ .

5. Deeply staining granules within the cell nucleus form the *chromatin network* ~~nucleolus~~.

6. The upper part of the dorsal body cavity is the .... *cranial cavity*

IV. LABELING

For each of the following illustrations, print the name or names of each labeled part on the numbered lines.

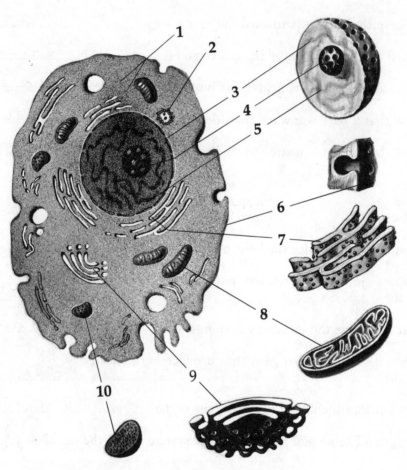

A typical cell.

1. *cytoplasm*
2. *centriole*
3. ~~nucleolus~~ *nucleus*
4. *nucleolus*
5. *chromatin network*
6. *cell membrane*
7. *endoplasmic reticulum*
8. *mitochondrion*
9. *Golgi body*
10. *lysosome*

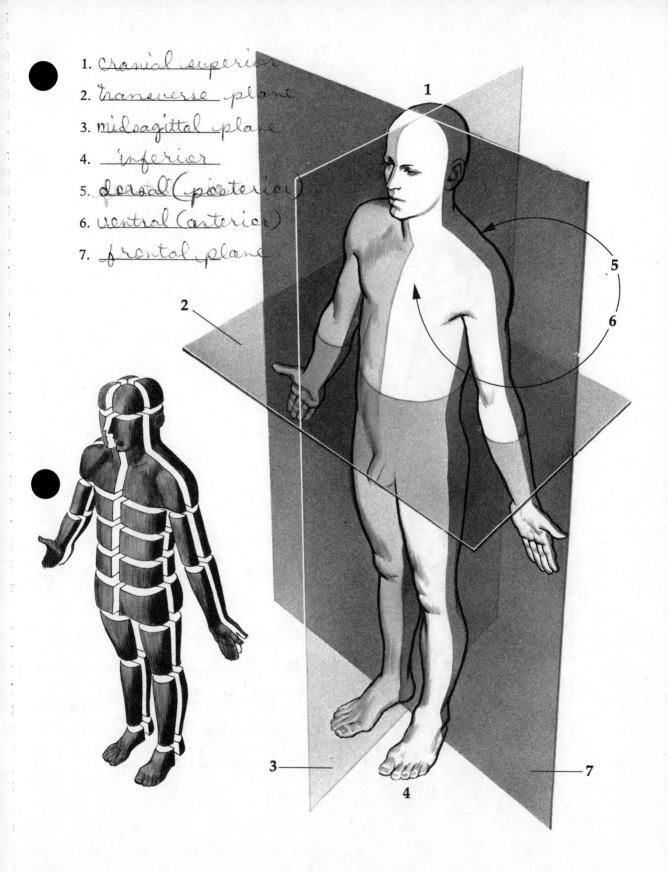

1. Cranial superior
2. transverse plane
3. midsagittal plane
4. inferior
5. dorsal (posterior)
6. ventral (anterior)
7. frontal plane

Body planes and directions.

13

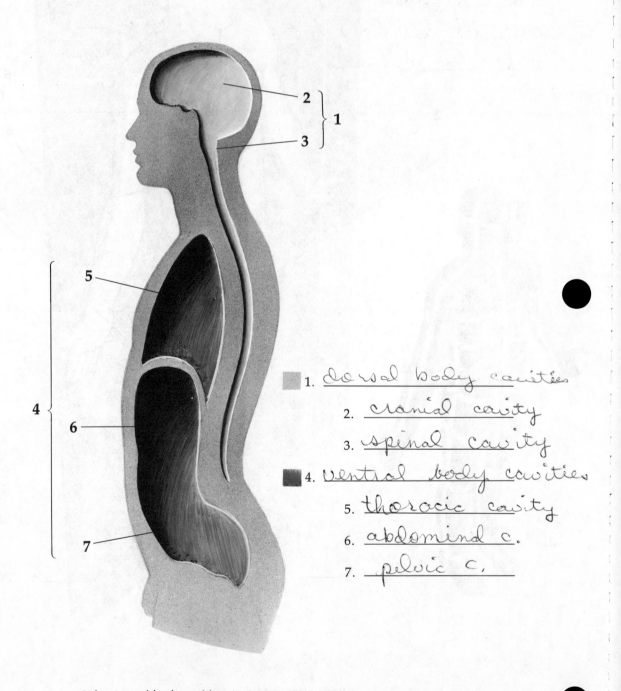

1. <u>dorsal body cavities</u>
2. <u>cranial cavity</u>
3. <u>spinal cavity</u>
4. <u>ventral body cavities</u>
5. <u>thoracic cavity</u>
6. <u>abdominal c.</u>
7. <u>pelvic c.</u>

Side view of body cavities.

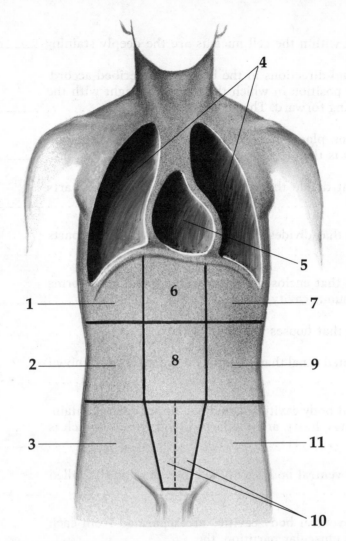

Front view of body cavities and the regions of the abdomen.

1. R. hypochondriac region
2. R. lumbar region
3. R. iliac region
4. pleural sacs for lungs
5. pericardial sac for heart
6. epigastric region
7. L. hypochondriac reg.
8. umbilical reg.
9. L. lumbar reg.
10. hypogastric reg.
11. L. iliac reg.

V. COMPLETION EXERCISE

Print the word or phrase that correctly completes the sentence.

1. The protoplasm enclosing the nucleus is called .... *cytoplasm*

2. Contained within the cell nucleus are the deeply staining _chromatin network_.

3. Regions and directions in the body are described according to the position in which the body is upright with the palms facing forward. This is called the .... _anatomic ~~central~~_.

4. The midline plane that divides the body into right and left halves is the .... _midsaggital_.

5. Planes that divide the body into upper and lower parts are called .... _transverse_.

6. The plane that divides the body into front and rear parts is the .... _frontal_.

7. The space that encloses the brain and spinal cord forms one continuous cavity, the .... _dorsal_.

8. The space that houses the brain is the .... _cranial cavity_.

9. The elongated canal that contains the spinal cord is known as the .... _spinal_.

10. The ventral body cavities include an upper space containing the lungs, heart, and the large blood vessels, which is called the .... _thoracic_.

11. The lower ventral body cavity is quite large and is called the .... _pelvic_.

12. The large ventral body cavities are separated from each other by a muscular partition, the .... _diaphragm_.

13. The large lower ventral body cavity is subdivided into nine regions including 3 nearer the midline. The uppermost of these midline areas is the .... _epigastrium_.

14. The standard metric measurement for volume, slightly greater than a quart, is called a .... _liter_.

VI. PRACTICAL APPLICATIONS

Study each discussion. Then print the appropriate word or phrase in the space provided.

Group A

1. The gallbladder is located just below the liver. The directional terms that best describe this relationship include _inferior_.

2. The kidneys are located behind the other abdominal organs. This relationship may be described as .... _dorsal_.

16

3. The tips of the fingers and toes are farthest from the region of origin of these digits so they are said to be the most _distal_.

4. The entrance area of the stomach is nearest the point of origin or beginning of the stomach so this part is said to be .... _proximal_.

5. The ears are located away from the midsagittal plane or toward the side so they are described as being .... _lateral_.

6. The head of the pancreas is nearer the midsagittal plane than its tail portion, so the head part is more .... _medial_.

7. The diaphragm is above the abdominal organs; it may be described as .... _superior_.

Group B

On the ward in which postoperative patients were being cared for you were asked to study certain cases and answer the following questions.

1. Mr. A had an appendectomy. The area of the abdomen in which the appendix is located is in the lower right side and is known as the .... _lateral_.

2. Mrs. B had a history of having gallstones. The operation to remove these stones involved the upper right part of the abdominal cavity, or the .... _superior_.

3. Miss C was injured in an automobile accident. In addition to a number of fractures she suffered a ruptured urinary bladder. The area in the lower midline part of the abdomen is the .... _lateral_.

4. Mr. B required an extensive exploratory operation that necessitated incision through the navel. This portion of the abdomen is the .... _proximal_.

5. Some of these patients were given aspirin for relief of pain. As with all drugs, the dosage is now measured by the metric system in grams or smaller units known as ...... _milligrams_.

VII. REFERENCES

Memmler, R. L., and Wood, D. L.: The Human Body in Health and Disease, ed. 4, pp. 1-9. Philadelphia, Lippincott, 1977.

Chaffee, E. E., and Greisheimer, E. M.: Basic Physiology and Anatomy, ed. 3, pp. 1-14. Philadelphia, Lippincott, 1974.

Crouch, J. E., and McClintic, J. R.: Human Anatomy and Physiology, p. 6. New York, Wiley, 1971.

Greisheimer, E. M., and Wiedeman, M. P.: Physiology and Anatomy, ed. 9, pp. 2-11. Philadelphia, Lippincott, 1972.

CHAPTER TWO

Disease and Disease-Producing Organisms

I. OVERVIEW

Disease may be defined as an impairment or other change from the normal state which prevents some of the tissues and organs from carrying on their required function. The causes are many and varied. Among them are congenital disorders or birth defects (which may be hereditary or acquired within the uterus), disorders resulting from *environmental factors* (such as a chemical agent, or lack of sunshine) and those due to *abnormal cell growth* (neoplasia). *Predisposing causes* are another important group of factors that play a part in the development of disease. An understanding of disease incorporates a study of the body including its structure (anatomy) and its functions (physiology) under normal and abnormal conditions (pathology). This understanding is aided by use of *disease terminology*. *Infection*, or invasion of the body by *disease-producing microorganisms*, including bacteria, fungi, viruses and protozoa, is the most important cause of disease in human beings. A second major cause of human disease is *infestation*, a special type of infection due to *parasitic worms*. Public health has been vastly improved through *laboratory identification* of pathogens and the application of *chemotherapeutic* and *aseptic* methods to prevent or control their spread.

II. TOPICS FOR REVIEW

1. characteristics of pathogens
2. nutritional factors in disease development
3. physical and chemical agents that cause injury and disease
4. other environmental factors that contribute to disease production
5. disease terminology
6. basis of infection
7. methods of destroying pathogens or of inhibiting their growth

III. MATCHING EXERCISES

Matching only within each group, print the answer in the space provided.

Group A

hereditary	congenital	therapy
symptoms	acute	chronic
pathogens	malnutrition	

1. A lack of essential substances such as vitamins in one's diet is known as .... *malnutrition*

2. Living organisms that cause many types of illness throughout the world are called .... *pathogens*

3. Any defect or abnormality present at birth is said to be *congenital*.

4. A disorder that is passed on to the infant by way of the patient's reproductive cells is referred to as .... *hereditary*.

5. A relatively severe disorder of short duration is said to be *acute*.

6. Diseases that persist over a long period of time are said to be .... *chronic*.

7. Changes in body function that are experienced by the patient are called .... *symptoms*.

8. A course of treatment that is prescribed by the physician is called .... *therapy*.

Group B

viruses	rickettsias	staphylococci
fungi	bacilli	streptococci
tuberculosis and leprosy	amebae	chlamydias

1. Rod-shaped bacteria are called .... *bacilli*.

2. Dot-shaped bacteria found in chains are known as .... *streptococci*.

3. Dot-shaped cells that appear in clumps are called .... *staphylococci*

4. Small obligate parasites that cause typhus fever and certain other febrile diseases are .... *rickettsias* ~~tuberculosis & leprosy~~

5. Mycotic infections are caused by simple plants, the ..... *fungi*

6. Certain organisms are classified with bacteria, though they are smaller. These organisms are the cause of trachoma (an eye disease) and parrot fever. They are known as .... *chlamydias*.

7. The smallest known living things are believed to be the _viruses_.

8. Among the protozoa are those disease-producing microscopic animals called the .... _amebae_.

9. Among important acid-fast organisms are those that cause ~~rickettsias~~ _TB + lep_.

Group C

1. The <u>process</u> that kills every living organism on an object is _asepsis_.

2. A <u>condition</u> in which no disease-causing organisms are present is called .... _sterilization_

3. The process of determining the nature of an illness is called making a .... _diagnosis_ ~~etiology~~

4. A group of signs or symptoms that occur together forms a _syndrome_.

5. A disease present at the same time in many people living in the same area is said to be .... _epidemic_.

6. If a disease is characteristically present continuously in a given area it is said to be .... _endemic_ ~~syndrome~~

7. A disorder without known cause, or self-originating, is said to be .... _idiopathic_.

8. The study of the <u>cause</u> of a disorder is said to be its ..... _etiology_

9. The range of occurrence of a disease and its tendency to affect particular groups of persons is said to be its ...... _incidence_.

Group D

1. The study of one-celled animals is .... _protozoology_.

2. The study of organisms that live on or within other organisms at the expense of those organisms is .... _parasitology_

3. The study that deals with the activities or functions of a living organism and its parts is .... _physiology_.

4. The study of one-celled plants or plantlike organisms is known as .... _bacteriology_

5. The study which includes both microscopic plants and animals is .... _microbiology_.

6. The science that deals with the structure and relationships of parts of the body is .... _Anatomy_.

7. The study that deals with the nature of disease and includes the changes caused by disease is .... _Pathology_.

8. A disease that can be transmitted from one person to another is .... _communicable_

9. Infections caused by fungi are referred to as .... _mycotic_.

Group E

infection	spore	pasteurization
botulism	dust	gram-positive
acid-fast	chemotherapy	

1. The process of heating milk to 145°F. for 30 minutes and then allowing it to cool rapidly before it is bottled is .... _pasteurization_.

2. The stage in which pathogenic bacteria develop a protective coat is the .... _spore_.

3. A deadly type of food poisoning due to a rod-shaped bacillus is known as .... _botulism_.

4. A most effective carrier of microbes present in the atmosphere is .... _dust_.

5. Invasion of the body by pathogenic microorganisms is called an .... _infection_.

6. The organisms that cause tuberculosis and leprosy retain color (stain) after application of an acid. Such organisms are said to be .... _acid-fast_.

7. Microorganisms that retain the bluish dye after they have been stained and treated with iodine and a solvent are said to be .... _gram-positive_.

8. The treatment of disease by administration of chemical substances that act selectively to destroy bacteria or to inhibit their growth is called .... _chemotherapy_.

IV. LABELING

For each of the following illustrations, print the name or names of each labeled part on the numbered lines.

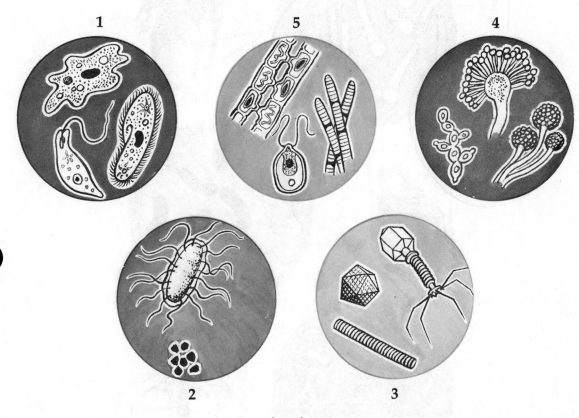

Examples of protists.

1. _protozoa_

2. _bacteria_

3. _viruses_

4. _fungi_

5. _algae_

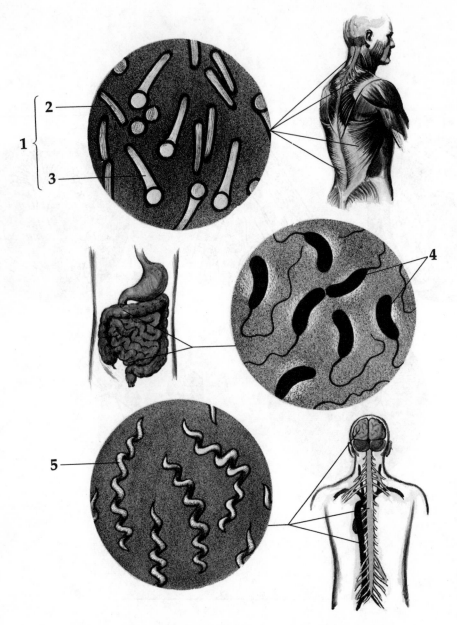

Bacteria of the elongated type.

1. <u>tetanus (lockjaw)</u>   4. <u>asiatic cholera</u>

2. <u>vegetative form</u>   5. <u>syphilis</u>

3. <u>spore form</u>

24

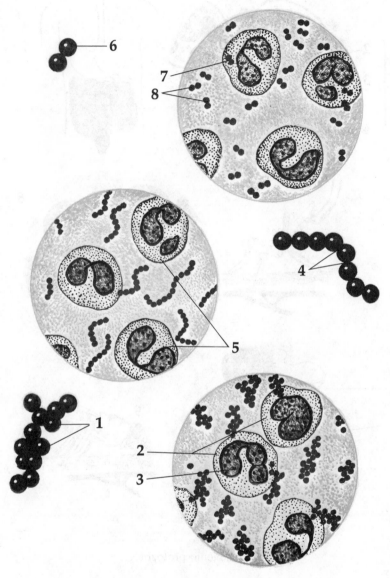

Bacteria of the spherical type.

1. ~~staphylococci~~

2. ~~pus cells~~

3. nucleus of leukocyte

4. streptococci

5. leukocytes

6. diplococci

7. intracellular cocci

8. extracellular cocci

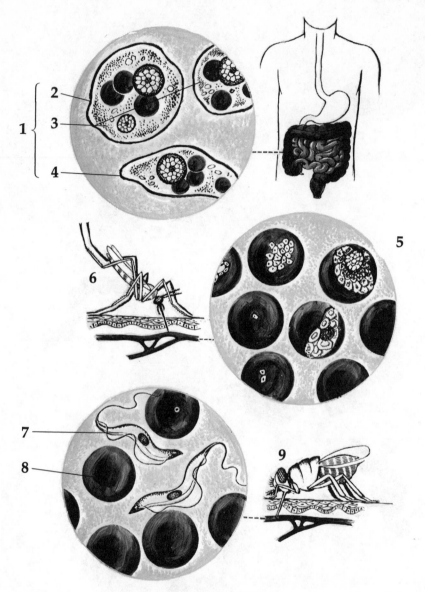

Pathogenic protozoa.

1. _Amebic dysentery_   6. _Mosquito_

2. _protective wall of cyst_   7. _Trypanosoma gambiense_

3. _inactive forms (cysts)_   8. _erythrocyte_

4. _active forms of ameba_   9. _tsetse fly_

5. _Malaria_

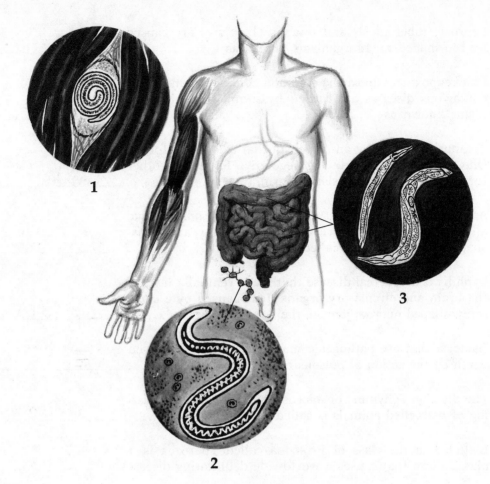

Common parasitic worms.

1. _____trichina_____   3. _____ascaris_____

2. _____filaria_____

V. COMPLETION EXERCISE

Print the word or phrase that correctly completes the sentence.

1. The most prevalent contagious and reportable disease at the present time involves the mucous membranes of the reproductive tract. It is called .... _syphilis_.

2. Boils, carbuncles and impetigo are among the infections caused by ubiquitous (present everywhere) organisms called .... _staphlococci_.

3. Blood poisoning (septicemia), scarlet fever, rheumatic fever, and other disorders are caused by a group of micro-organisms called .... _strepococci_ ~~_pathogens_~~.

27

4. Leprosy, tuberculosis and tetanus (lockjaw) are caused by rod-shaped microorganisms known as .... _bacilli_ .

5. Chickenpox, smallpox, the common cold and many other contagious diseases are caused by submicroscopic organisms known as .... _viruses_ .

6. Tetanus is caused by an organism that exists in 2 forms. One of these is the growing vegetative form. The other form is a resting and resistant form, the .... _spore_ .

7. Asiatic cholera is a dread disease found in India, China, and other Asiatic countries. It is caused by a comma-shaped microorganism called the .... _vibrio_ .

8. Syphilis is a venereal disease that may eventually involve the brain and circulatory organs. It is caused by a cork-screw-shaped microorganism, the .... _spirillum_ .

9. Bacteria that are pathogenic may cause injury and even death by the action of poisons referred to as .... ~~spirochaetes~~ ~~toxins~~ _toXINS_ .

10. The division (phylum) of microscopic organisms consisting of one-celled animals is called .... _protozoa_ .

11. Included in the class of Protozoa called Sporozoa is a plasmodium that causes a worldwide debilitating disease _malaria_ .

12. Among the most persistent parasitic infections is that due to the small seatworm or pinworm, the scientific name for which is .... _Enterobius vermicularis_

13. A small roundworm that may become enclosed in a cyst or sac inside the muscles of the pig and be transmitted to man via improperly cooked pork is known as the ..... _(tapeworm)_ _trichinella_ .

14. The flatworm composed of many segments (proglottids) may grow to a length of 50 feet—hence the name .... _tapeworms_ .

VI. PRACTICAL APPLICATIONS

Study each discussion. Then print the appropriate word or phrase in the space provided.

1. Mrs. K brought her young baby in for follow-up examination of his condition of harelip. Harelip is an example of a birth defect that is acquired by the fetus during development in the mother's uterus. Such defects, along with those that are inherited, are all said to be .... _congenital_ .

2. Mrs. J brought her son, age 3, to the clinic because he had suffered undue blood loss from a small cut on his hand. It was found that the child had an inherited blood disorder called .... *hemophilia.*

3. Mr. N, age 81, showed evidence of tissue deterioration. This wear and tear is often described as .... *degeneration*

4. Miss C required further study of an obscure disorder for which no cause had been found. Such a disease is often referred to as .... *idiopathic*.

5. Mr. S needed prophylactic (preventive) treatment because he had received several puncture wounds while repairing an old building. He was given an injection to prevent the development of lockjaw. The scientific name for lockjaw is .... *tetnus.*

6. Miss D attended the clinic for treatment of a very sore throat. The diagnosis was streptococcal ("strep") sore throat. The full name for the bacteria that cause this infection is .... *strepolococci*

7. Mrs. A brought her young daughter to the clinic for examination of boils (furuncles) that had been appearing on the child's neck with disturbing frequency. Boils are often due to dot-shaped microorganisms that resemble bunches of grapes when examined under the microscope. These organisms are called .... *stofalococci.*

8. Mr. E came to the clinic because he had suffered from bouts of diarrhea ever since returning from a camping trip that took him into several countries. One common cause of intestinal disorders is a dysentery caused by a one-celled protozoan called the .... *amebae.*

VII. REFERENCES

Memmler, R. L., and Wood, D. L.: The Human Body in Health and Disease, ed. 4, pp. 11-26, Appendix, Tables 1-4. Philadelphia, Lippincott, 1977.

Frobisher, M.: Fundamentals of Microbiology, ed. 9, pp. 149-152, 188-190, 310-316, 323-329, 551-568, 610-617, 632-646. Philadelphia, Saunders, 1974.

Nason, A., and Dehaan, R. L.: The Biological World, pp. 146-154. New York, Wiley, 1973.

Salle, A.: Fundamental Principles of Bacteriology, ed. 7, pp. 72-74, 586-593, 1005-1035. Los Angeles, McGraw-Hill, 1973.

Stonehouse, B.: The Way Your Body Works, pp. 72-75. London, Beasley, 1974.

CHAPTER THREE

Chemistry, Matter and Life

I. OVERVIEW

Chemistry is the *physical science* that is concerned with the atoms, elements, molecules, compounds and mixtures that are the fundamental units of matter. Though exceedingly small particles, atoms possess a definite structure: the *nucleus* contains *protons* and *neutrons*, and surrounding the nucleus are the *electrons*. An *element* is a substance consisting of just one type of atom. Union of two or more atoms produces a *molecule*; the atoms may be alike (such as the oxygen molecule) or different (sodium chloride, for example) and in the latter case the substance is called a *compound*. To go a step further, a combination of compounds, each of which retains its separate properties, is a *mixture* (salt water is one example). Through the process of *electron transfer*, which goes on ceaselessly in all living things, chemical compounds are constantly being formed, altered, broken down and recombined into other substances. Hydrogen, oxygen, carbon and nitrogen are the elements that constitute about 99 percent of protoplasm, while calcium, sodium, potassium, phosphorus, sulfur, chlorine and magnesium account for most of the remaining 1 percent. Proteins, carbohydrates and lipids are among the compounds formed from these elements, and are essential to proper growth and development of the human organism. To function effectively in their chosen field, all persons engaged in the health occupations should possess an understanding of these basic principles of chemistry.

II. TOPICS FOR REVIEW

1. atoms and elements
2. molecules, compounds and mixtures
3. ions and electrolytes
4. acids and bases
5. carbohydrates, proteins and lipids (fatlike substances)
6. mineral (inorganic) salts
7. main elements of protoplasm
8. the sol state and the gel state
9. proteins and amino acids

III. MATCHING EXERCISES

Matching only within each group, print the answer in the space provided.

Group A

chemicals	atoms	nucleus
chemistry	pharmacology	microbiology

1. The smallest complete units of matter are called .... _atoms_ .

2. The science that deals with the composition of all matter is _____ .

3. The study of all aspects of drugs is called .... _pharmacology_ _chemistry_ .

4. Aspirin, hexachlorophene and penicillin are classified as.. _chemicals_ .

5. The part of the atom containing most of its mass including protons and neutrons is the .... _____ .

6. The science that deals with the study of microorganisms is known as .... _____ .

Group B

radioactivity	isotopes	elements
molecule	compounds	mixture
neutrons	electrons	protons

1. The positively charged particles inside the atomic nucleus are .... _____ .

2. The noncharged particles within the atomic nucleus are .. _____ .

3. The negatively charged electric particles outside the atomic nucleus are the .... _____ .

4. Substances composed of one type of atom are called .... _____ .

5. Elements existing in forms that are alike in their chemical reactions but that differ in weight are called .... _____ .

6. The word that refers to emission (giving off) of rays from disintegrating isotopes is .... _____ .

7. The unit formed by the union of 2 or more atoms is the .. _molecule_ .

8. Substances that result from the union of 2 or more elements are known as .... _compounds_ .

9. The combination of various compounds that remain intact and retain their properties is designated a .... _____ .

32

Group C

carbohydrates	proteins	rays
cations	anions	electrolytes

1. Positively charged ions are called .... _____.

2. Negatively charged ions are known as .... _____.

\3. Compounds that form ions when in solution are called .. _electrolytes_.

4. Compounds of nitrogen, carbon, oxygen and hydrogen include the .... _____.

5. Compounds of carbon, hydrogen and oxygen (such as the simple sugars) are classified as .... _carbohydrates_.

6. Small particles given off by radioactive substances are referred to as .... _____.

Group D SKIP

sol state	gel state	protoplasm
iron	homeostasis	electron transfer
calcium ions	carbonate ions	pH

1. When a positively charged cation (such as a sodium ion) comes in contact with a negatively charged anion (such as a chlorine ion) a new compound is formed. This process is called .... _____.

2. The clotting of blood, normal muscle relaxation and bone tissue health all require the presence of .... _____.

3. In maintaining a normal stable balance between acidity and alkalinity certain ions are necessary, notably .... _____.

4. The stable condition of body chemistry which is heavily dependent on certain ions is called .... _____.

5. The degree of acidity or hydrogen ion concentration is indicated by the .... _____.

6. Three light gaseous elements and one solid element combine to form 99% of the living material called .... _____.

7. Only traces of certain well-known elements are found in living material. One of these is .... _____.

8. The semisolid condition in which protoplasm may sometimes be found is called the .... _____.

9. The fluid condition in which protoplasm may be found at certain times is known as the .... _____.

Group E

chemical principles neutrons   different
radioactivity   decomposed   element
liquid   organic   amino acids

1. To understand body processes such as the digestion of food and respiration, the health worker needs to know .. _____.

2. A substance composed of one type of atom only is a(n) _____.

3. A compound is formed by the union of 2 or more atoms which are .... _____.

4. The cobalt bomb, used in the treatment of cancer, utilizes the property of certain elements which is called .... _____.

\5. Essential for proper nutrition because they are the building blocks of proteins are the .... *amino acids*.

6. Elements occur as gases, solids or liquids. Mercury is an example of a .... _____.

7. The greater weight of an isotope is due to the presence of a larger number of .... _____.

8. Elements cannot be changed into something else by physical or chemical methods. That is, they cannot be .... _____.

9. Most of the compounds that are found in living organisms are classified as .... _____.

IV. LABELING

For each of the following illustrations, print the name or names of each labeled part on the numbered lines.

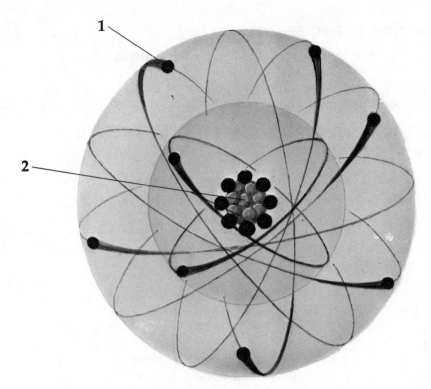

Oxygen atom.

1. _electron_

2. _central nucleus_

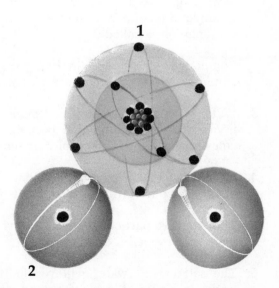

1. _oxygen atom_

2. _hydrogen atom_

Molecule of water.

V. COMPLETION EXERCISE

Print the word or phrase that correctly completes the sentence.

1. The 4 elements that make up about 99 percent of protoplasm are carbon, hydrogen and oxygen plus .... _____.

2. The fatlike substances formed from carbon, hydrogen and
oxygen are the .... _____.

3. The sum of all the physical and chemical processes that
enable protoplasm to produce energy and build protein is
called .... _____.

4. Those amino acids that cannot be synthesized in adequate *(essential)*
amounts by the body cells must be supplied in the diet. *dispensable*
They are ....

5. Proteins that contain all the necessary amino acids are *NON- essential*
described as .... *in- dispensable*

6. A mixture of protein foods in a diet provides enough
amino acids because one food supplements another.
Beans, peas and peanuts are examples of foods that contain proteins described as .... _____.

VI. PRACTICAL APPLICATIONS

Study each discussion. Then print the appropriate word or phrase in the space
provided.

Group A

1. Most people keep a shaker of salt on the table. Salt is an
example of a combination of 2 different elements. Such a
combination is called a .... _____.

2. Compounds first found in living organisms, as for example starch in potatoes, are classified as .... _____.

3. Water is certainly the most common compound known. It
is classified as a compound because it is made of molecules formed by the union of 2 or more .... _____.

4. An element that is part of the air we breathe also is part
of the water we drink. This element is .... _____.

5. The smallest particle of salt obtainable that would still
have the properties of salt is the .... _____.

6. The salt in salt water will regain its properties if the
water is boiled. Since water and salt do not combine
chemically, this solution is an example of a(n) .... _____.

7. If we could remove a single electron from a sodium atom, or could add a single electron to a sodium atom, the result would be an atom with either a positive or a negative charge, known as a(n) .... _____.

8. Numerous essential body activities are possible due to the property of certain compounds to form ions when in solution. Such compounds are called .... *electrolyte* ~~organic~~ .

9. Inorganic salts such as potassium chloride must be present in exactly the right quantities for proper body function. Another term for these electrolytes is .... _____.

Group B)Acc

Review assigned clinic charts to learn about the kinds of tests performed. You will find that most of these studies are based on principles of chemistry and physics.

1. Mr. B complained of shortness of breath. Several studies were done including a visible tracing of the electric currents produced by his heart muscle. Such a record is called an .... *EKG or ECG* *electrocardiogram*

2. Joan, age 4, was brought to the clinic by her mother because she experienced attacks of fainting and unconsciousness. As an aid in diagnosis, a graphic record of her brain's electric current was obtained. This brain wave record is called an .... *EEG* *electroencephalo-gram*

3. A routine test done on Miss J showed glucose in her urine —an abnormal finding. Glucose is one of a group of compounds found in certain foods and classified as .... *carbohydrate* *sugar* .

4. Mr. K's urinalysis showed the presence of albumin. Albumin is an example of compounds found in protoplasm that contain nitrogen, carbon, hydrogen and oxygen. These compounds are classified as .... *protein* ~~crossed out~~ ~~crossed out~~ .

VII. REFERENCES

Memmler, R. L., and Wood, D. L.: The Human Body in Health and Disease, ed. 4, pp. 27-33. Philadelphia, Lippincott, 1977.

Chaffee, E. E., and Greisheimer, E. M.: Basic Physiology and Anatomy, ed. 3, pp. 16-18. Philadelphia, Lippincott, 1974.

Frobisher, M.: Fundamentals of Microbiology, ed. 9, pp. 59-64. Philadelphia, Saunders, 1974.

Greisheimer, E. M., and Wiedeman, M. R.: Physiology and Anatomy, ed. 9, pp. 41-52. Philadelphia, Lippincott, 1972.

Griffiths, M.: Introduction to Human Physiology, pp. 20-33. New York, Macmillan, 1974.

CHAPTER FOUR

Cells, Tissues and Tumors

I. OVERVIEW

While the organization of matter into atoms and molecules involves submicroscopic structures, the composition of protoplasm includes microscopic and larger parts beginning with *cells* and their *organelles*. Whether a cell exists alone or as one unit of a structure, its work goes on ceaselessly through its *organelles* ("little organs"). Among these are the *mitochondria*, which contain the catalytic *enzymes*. Through *mitosis*, new cells are constantly being formed. They receive nourishment, generate heat and energy, get rid of waste products and discharge secretions by means of the *semipermeable* cell wall; this entire process is known as *metabolism*.

The tissues are composed of specialized groups of cells having a common purpose, and are classified into four main groups: *epithelium, connective tissue, nerve tissue* and *muscle tissue.* An understanding of the structure and function of tissues is especially important in relation to the study of *tumors*, since various tumors arise most frequently in specific types of tissue.

If you learn tissue classification and try to understand differences in function you will better understand both normal and abnormal structure as you proceed.

II. TOPICS FOR REVIEW

1. cell organelles including mitochondria
2. DNA, RNA and enzymes
3. cell division or mitosis
4. the physical processes of diffusion, osmosis and filtration
5. energy and nonenergy foods
6. characteristics of tissues in general
7. the 4 main kinds of tissues and their anatomical characteristics
8. function of epithelium
9. function of connective tissue
10. function of nerve tissue
11. function of muscle tissue
12. origin of tumors
13. characteristics of benign tumors
14. characteristics and types of malignant tumors

III. MATCHING EXERCISES

Matching only within each group, print the answer in the space provided.

Group A

energy	glucose	amino acids
osmosis	diffusion	filtration
chromosomes	organelles	enzymes

1. Certain complex proteins that act as catalytic agents are classified as .... *enzymes* .

2. The microscopic structures that are present in practically all living cells and that regulate a variety of functions within the cells are the .... *Organelles* .

3. Molecules move from an area of relatively high concentration to an area of lower concentration in the process of *diffusion* .

4. Rod-shaped structures that are deeply staining and are found only during cell division are known as .... *chromosomes* .

5. The passage of a solvent through a semipermeable membrane from an area of lower concentration to one that is higher is the process of .... *osmosis* .

6. The passage of water with its dissolved substances through a membrane as a result of a greater mechanical force on one side is the process called .... *filtration* .

7. During digestion carbohydrates are changed to a simple sugar called .... *glucose* .

8. The end products of protein digestion are .... *amino acids* .

9. Metabolism of foods produces capacity for action, that is, *energy* .

Group B

resting stage	adipose	secretions
tendons	cilia	mitochondria
cartilage	bone	inheritance

1. The chromatin material of the cell nucleus exists as granules during the cell's usual .... *resting stage* .

2. The genes, which are contained in the chromosomes, regulate .... *inheritance* .

3. Enzymes stimulate chemical activities within the cell. These activities go on inside the organelles called ...... *mitochondria*

4. An important function of epithelium is the production of _secretions_.

5. The tiny hairlike protoplasmic extensions that project from epithelium which help to prevent lung damage by keeping the airways clear are the .... _cilia_ .

6. The storing up of fat, heat insulation and padding of various structures are functions of the type of connective tissue called .... _adipose_ .

7. The strong, cablelike connective tissue cords that connect muscles to bones are .... ~~cartilage~~ _tendon_ .

8. One of the hard connective tissues that has the important function of acting as a shock absorber and as a bearing surface to reduce friction between moving parts is ...... _cartilage_ .

9. Osseous tissue is similar to cartilage in its cellular structure. Cartilage may gradually become impregnated with calcium salts to form .... _bones_ .

Group C

catalysts	tissue fluid	edema
nerve cells	semipermeable	dehydration
water	tissues	metabolism
physical and chemical changes	DNA	RNA

1. Substances that increase the speed of chemical reactions without being changed themselves are .... _catalystes_ .

2. Within the cell, the production of heat and energy, new protoplasm and waste products is known as .... _metabolism_ .

3. Organized groups of cells that are of the same type and have a common purpose form .... _tissues_ .

4. The basic structural units of nerve tissue are the .... _nerve cells_ .

5. From 60 percent to 99 percent of body tissues are made up of the abundant compound .... _water_ .

6. Bathing the tissues is a slightly salty solution, the ...... _tissue fluid_

7. In conditions of excessive fluid loss the tissues suffer from _dehydration_

8. The chief component of the chromosomes of the nucleus is a complex molecule called .... _DNA_ .

41

9. Instructions given by the DNA molecule are carried from the nucleus of the cell to other parts of the cell and to all parts of the body by another complex molecule called ... _RNA_.

10. A puffiness of tissues due to an abnormal accumulation of fluid is found in the condition called .... _edema_.

11. Metabolism is the term that describes a combination of activities within the cell including all the .... _physical + chemical changes_.

12. The cell wall is selective, that is, it permits some substances to enter the cell but prevents passage of others. Therefore, it is said to be .... _semipermeable_.

Group D

myocardium	fibers	connective tissue
cell division	neurilemma	myelin
voluntary muscle	spasm	visceral muscle

1. Areolar, adipose and osseous tissue all act as the body's supporting fabric and are therefore classified as .... _connective tissue_ ~~neurilemma~~.

2. The basic structural unit of nerve tissue, the neuron, consists of a nerve cell body plus small branches, which are called .... _fibers_.

3. The ability of certain nerves to repair themselves is due to the presence of .... _neurilemma_ ~~connective tissue~~.

4. Like telephone wires, nerve fibers are encased in a protective covering, or sheath. This fatty insulating material is called .... _myelin_.

5. The thickest layer of the heart wall is formed by cardiac muscle or .... _myocardium_.

6. Certain diseases are characterized by abnormal muscle contractions. A single sudden violent contraction is classified as a .... _spasm_.

7. Muscle tissue is classified into 3 types. That which forms the walls of the organs within the ventral body cavities is called .... _visceral muscle_.

8. Skeletal muscle provides for the movement of the body. It is therefore described as .... _voluntary muscle_.

9. Even after maximum growth is attained there is still a continual (though slower) process of cell production by means of .... _cell division_.

Group E

suture	benign	neurilemma
epithelium	malignant	tumor
proud flesh	connective tissue	myelin

1. The main tissue of many protective coverings and of the linings of the respiratory and digestive tracts, among others, is called .... *epithelium.*

2. Repair of damaged nerve and muscle tissue is accomplished by the growth of .... *connective tissue*

3. Forming a sheath around some nerve fibers is a fatlike material that acts as insulation. It is called .... *myelin.*

4. During the process of repairing a large wound an excessive growth of blood vessels may cause the appearance of so-called granulation tissue. The more common name for these rounded fleshy masses is .... *proud flesh* ~~*Tumor*~~.

5. Repeated injury to a single area may trigger the development of abnormal tissue growth at this point. Such a growth is given the general name of .... *tumor* ~~*proud flesh*~~

6. The size of a scar following the healing of a clean wound may best be reduced by bringing the edges together with a .... *suture.*

7. Nerves located outside the central nervous system (peripheral) may repair themselves with the help of a thin coating membrane called .... *neurilemma*

8. New growths that spread and grow rapidly, often causing death, are said to be .... *malignant.*

9. A growth that does not spread and is usually confined to a local area is said to be innocent or .... *benign.*

Group F

sarcoma	nevus	angioma
carcinoma	lipoma	osteoma
papilloma	myoma	glioma

1. A projecting mass of epithelium is called a wart or ...... *papilloma.*

2. A benign connective tissue tumor that originates in adipose tissue is called a .... *lipoma.*

3. The general term for a tumor composed of blood and lymph vessels is .... *angioma.*

43

4. A tumor that originates from the special connective tissue of the central nervous system is a .... *glioma*.

5. A benign connective tissue tumor that originates in a bone is called an .... *osteoma*.

6. A better term for a mole or other small circumscribed tumor of the skin is .... *nevus*.

7. The most common type of cancer is one that originates in epithelium. It is called a .... *carcinoma*.

8. Cancers that originate in connective tissue usually spread via the blood stream. This type of cancer is called a ..... *sarcoma*.

9. The fibroid originates in the uterus. This innocent tumor is correctly classified as a .... *myoma*.

IV. LABELING

For each of the following illustrations, print the name or names of each labeled part on the numbered lines.

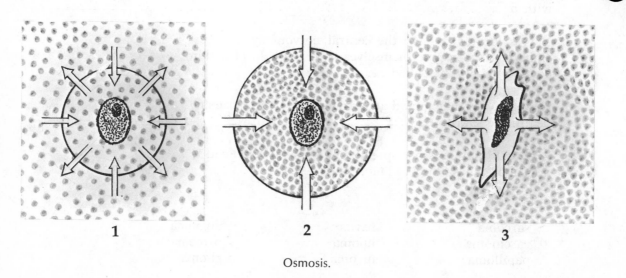

1   2   3

Osmosis.

1. _____

2. _____

3. _____

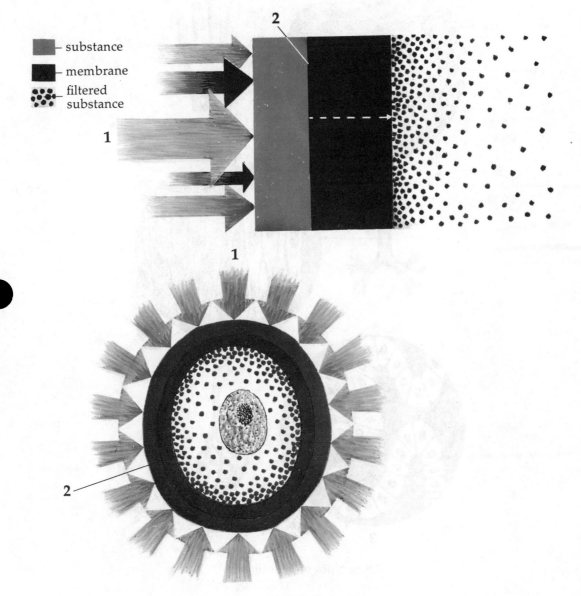

substance

membrane

filtered
substance

Filtration.

1. _____

2. _____

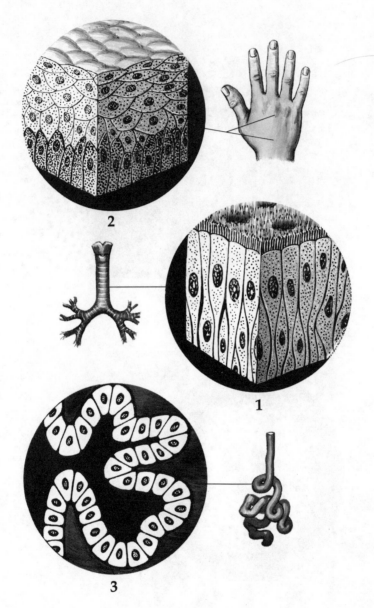

Three types of epithelium.

1. _____

2. _____

3. _____

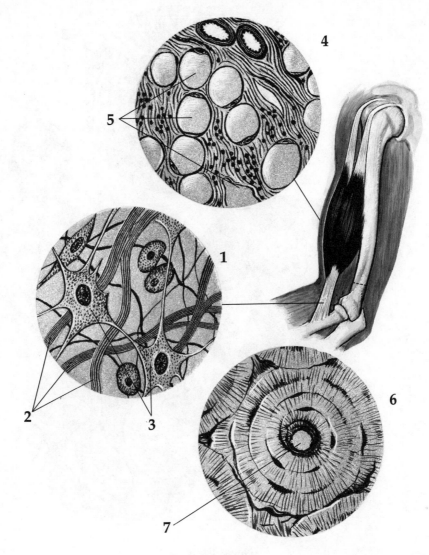

Connective tissue.

1. _____

2. _____

3. _____

4. _____

5. _____

6. _____

7. _____

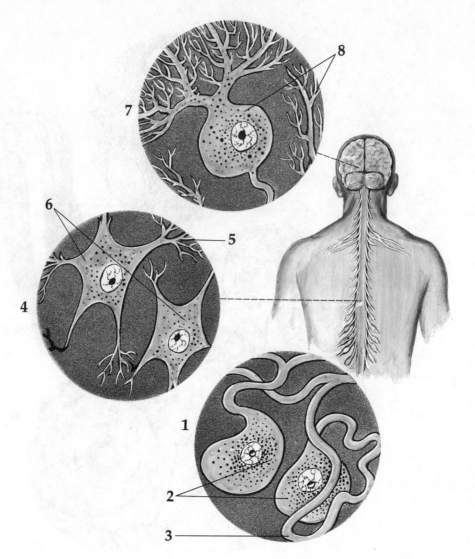

Nerve tissue.

1. _____

2. _____

3. _____

4. _____

5. _____

6. _____

7. _____

8. _____

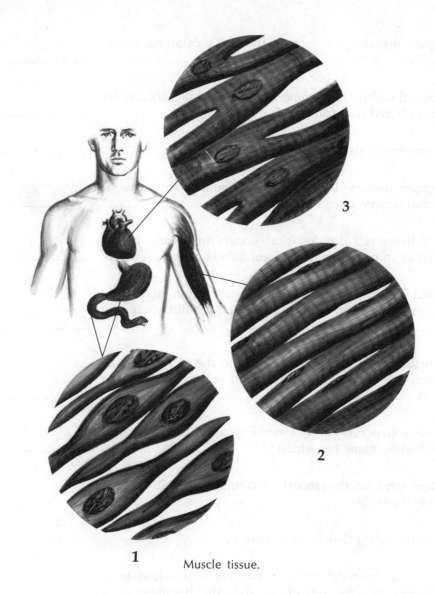

Muscle tissue.

1. _____

2. _____

3. _____

V. COMPLETION EXERCISE

Print the word or phrase that correctly completes the sentence.

1. The name of the fundamental building block that living
   things are made of is .... _____*cell*_____.

49

2. Complex proteins that act as the cell's catalytic agents are called .... *enzymes* .

3. Contained within the chromosomes are the factors responsible for hereditary traits, the .... *genes* .

4. The connective tissue of the brain is .... *neurilemma*

5. Malignant tumors derived from the connective tissue of the central nervous system are .... *gliema* .

6. The indirect cell division that occurs in human beings as well as in many plants and animals is called .... *mytosis* .

7. Normal saline solution has a concentration about the same as that of the body fluids. Such a solution is said to be .... *isotonic* .

8. The broad general term that refers to the sum of all the physical and chemical processes that occur in the human body is .... *metabolism* .

9. The basic structural unit of nerve tissue is the nerve cell, the scientific name for which is .... *neuron* .

10. Another term for the smooth involuntary muscle of most hollow organs is .... *visceral* .

11. The word ending that means tumor is .... *oma* .

12. The general term referring to cancers that originate in epithelium and are spread through the lymphatic system is .... *carcinoma* .

13. The malignant tumors that originate in connective tissue and that spread via the bloodstream are called .... *sarcoma* .

14. By far the most common type of cancer is .... *carcinoma* .

15. Among the forms of treatment of certain cancers is the use of drugs. This treatment is known as .... *chemotheraphy* .

VI. PRACTICAL APPLICATIONS

Study each discussion. Then print the appropriate word or phrase in the space provided.

Group A

Observations you might make while touring a hospital laboratory include the following.

1. The janitor in the laboratory was using a cleaning solution that contained ammonia. You will recall that this would cause ammonia molecules to spread throughout the room. This movement of molecules from an area of high concentration to other areas where concentration is low is called .... *diffusion* .

2. One of the laboratory technicians was trying to separate solid particles from a liquid mixture. He poured the mixture into a paper-lined funnel. The liquid flowed through the funnel while the solids remained behind on the paper. This process is called .... *filtration* .

3. A laboratory worker was carefully measuring certain mineral salts in order to prepare a normal saline solution. Normal saline is used to replace lost body fluids because the concentration is nearly the same as that inside the cells. Such a solution is said to be .... *isotonic* .

4. While doing a complete blood count a technician noted that some of the red blood cells had ruptured. The solutions used were tested to determine whether they were too dilute. Osmosis of water into a cell could be the cause of cell breakage. A too dilute solution is said to be ...... *hypotonic* .

5. A student was learning how to do blood smears. Upon examination of the blood with the micrscope he found that many red blood cells appeared shrunken. The explanation was that he was proceeding so slowly that the liquid part of the blood was evaporating, leaving a highly concentrated solution. Such a solution is described as being .... _____ .

Group B

A day in the tumor clinic involves observation of several patients. Among the situations you might encounter are the following.

1. Mr. B was concerned about a small growth on the left side of his face which had not cleared after many months. He thought it might be growing. The physician informed Mr. B that a biopsy was necessary in order to make a positive diagnosis. If the growth proved to be malignant it would be a .... *carcinoma* .

2. Mrs. C complained of an irritation in the area of a large dark mole on her right ankle. The doctor advised her to have the mole removed and examined under the microscope, because such a mole may be malignant. This type of cancer is called .... _____ .

3. Mr. K's problem involved multiple rounded growths located just under the skin of his right forearm. These were diagnosed as benign fatty tumors, or .... _Lipoma_ .

4. Miss G had noticed a small lump in her left breast. A biopsy was done in order to determine whether this was a malignant tumor of the ducts, which could be classified as ....

5. Baby K was under treatment for a large birthmark on his right cheek. This type of tumor, which is composed of blood or lymph vessels, is classified as an .... _angioma_ .

VII. REFERENCES

Memmler, R. L., and Wood, D. L.: The Human Body in Health and Disease, ed. 4, pp. 35-52. Philadelphia, Lippincott, 1977.

Chaffee, E. E., and Greisheimer, E. M.: Basic Physiology and Anatomy, ed. 3, pp. 18-42, 96-113, 163-173, 324, 327. Philadelphia, Lippincott, 1974.

Frobisher, M.: Fundamentals of Microbiology, ed. 9, pp. 250-251. Philadelphia, Saunders, 1974.

Gore, R.: "The Awesome Worlds Within a Cell," in the National Geographic, vol. 150, no. 3, September, 1976, pp. 355-395.

Nason, A., and Dehaan, R. L.: The Biological World, pp. 111-136. New York, Wiley, 1973.

Stonehouse, B.: The Way Your Body Works, pp. 10-13. London, Beasley, 1974.

CHAPTER FIVE

Membranes

I. OVERVIEW

The simplest combination of tissues is the *membrane*—the thin sheet of material that separates 2 groups of substances. In the group of *epithelial* membranes are 2 sub-groups: the *mucous* membranes lining tubes and other spaces that open to the outside of the body, and the *serous* membranes that line closed cavities inside the body. The second group of membranes, *connective tissue* membranes, is also divided into 2 sub-groups: the *fascial* membranes anchor and support the organs, and the *skeletal* membranes cover bone and cartilage.

Many diseases directly affect membranes. Pleurisy, for example, is an inflammation of the membrane lining the chest cavity. Then, too, membranes can serve as travel routes for disease, and can act to wall off an infection and prevent it from moving into another area.

As you learn about the various membranes you will find it helpful to think of structure and function together.

II. TOPICS FOR REVIEW

1. definitions: types of membranes
2. main categories of membranes
   a. epithelial membrane
   b. connective tissue membrane
3. subgroups of epithelial membranes
   a. characteristics and function of mucous membranes; examples
   b. characteristics and function of serous membranes; examples
4. subgroups of connective tissue membranes
   a. characteristics and function of fascial membranes; examples
   b. characteristics and function of skeletal membranes; examples
5. membranes and disease
   a. direct effects of disease on membranes
   b. how membranes serve to carry disease or to confine it

III. MATCHING EXERCISES

Matching only within each group, print the answer in the space provided.

Group A

pleurae	cell wall	serous membranes
pericardium	membrane	mucous membranes
peritoneum	lubricants	fascial membranes

1. Any thin sheet of material that separates 2 or more groups of substances is classified as a .... *membrane*.

2. The membrane that permits certain substances to enter the cell and certain substances to pass out is the .... *cell wall*.

3. The membranes that line the so-called closed cavities within the body are .... *serous membrane*

4. The tough membranes composed entirely of connective tissue which serve to anchor and support organs are the *fascial membrane*

5. The linings of tubes and spaces that are connected with the outside are largely epithelial. They are .... *Mucous Membrane*

6. The membranes that form the 2 separate sacs for the lungs are known as the .... *pleurae*.

7. The special sac that encloses the heart is known as the ... *pericardium*

8. The serous membrane of the abdominal cavity is the largest of its kind and it is called the .... *peritoneum*.

9. An important function of most epithelial membranes is to produce fluids that serve as .... *lubricants*.

Group B

superficial fascia	periosteum	synovial membranes
parietal layer	perichondrium	mucous membranes
mesothelium	capsules	

1. Membranous connective tissue envelopes that enclose organs are called .... *capsules*.

2. The tough connective tissue membrane that serves as bone covering is the .... *periosteum*.

3. Covering cartilage is a membrane similar to that covering bone. It is called .... *perichondrium*

4. Secretions produced by the linings of joint cavities act as lubricants to reduce friction between the ends of bones. These linings are .... *synovial*.

5. The linings of the various parts of the respiratory tract are all .... *mucous membranes*

6. The tissue that underlies the skin is known as the ... *superficial fascia*

7. The part of a serous membrane that is attached to the wall of a cavity or sac is the .... *parietal layer*.

8. Movements of organs occur with a minimum of friction because of the presence of a type of epithelium called .... *mesothelium*.

IV. LABELING

Print the name or names of each labeled part on the numbered lines.

1. *Greater Peritoneal*
2. *Lesser Peritoneal*
3. *liver*
4. *stomach*
5. *mesocolon*
6. *omentum*
7. *transverse colon*
8. *Sm. Intestines*
9. *uterus*
10. *urinary bladder*
11. *pancreas*
12. *duodenum*
13. *Retroperitoneal space*
14. *mesentery*
15. *cul-de-sac*
16. *rectum*

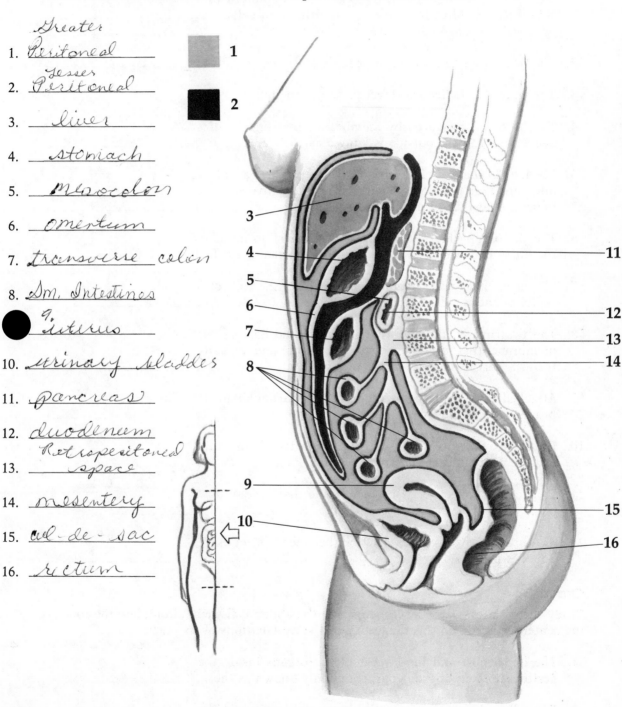

Abdominal cavity showing peritoneum.

V. COMPLETION EXERCISE

Group A
Print the word or phrase that correctly completes the sentence.

1. We have been referring throughout this chapter to thin skinlike layers that are made of a multitude of cells. The precise 2-word name for these layers is .... *tissue membrane*

2. The noun used to denote a mucous membrane is .... _____.

3. The noun that indicates a layer of serous membrane is ... _____.

4. The lubricant produced by membranes that line cavities connected with the outside is known as .... *mucous*.

5. The membranes that line closed cavities produce a watery lubricant that resembles a serum. The adjective used to indicate that type of membrane is .... _____.

6. The microscopic hairlike protoplasmic extensions found in the lining of most parts of the respiratory tract are called *cilia*.

7. The layer of a serous membrane that lines the wall of a cavity or sac is called the .... *periotal*.

8. The descriptive adjective that indicates the part of the membrane that is attached to the organ and actually forms a part of the wall of the organ is .... *visceral*.

9. Attached to the surface of the lung is a particular membrane that has the 2-word name .... *visceral pleural*.

10. Envelopes that are made of connective tissue and that enclose many internal organs are called .... *capsules*.

11. The tough connective tissue membrane that covers most parts of all bones is given the name .... *periosteum*.

12. A lubricant that reduces friction between the ends of bones is produced by the .... *synovial membrane*.

Group B
Study the diagram of the peritoneum and the abdominal cavity. Complete the following sentences based on this diagram and the explanations of its parts.

1. The uppermost and the largest of the organs inside the peritoneal cavity is the hepar, commonly known as the .. *liver*.

2. Extending downward from the stomach is an apronlike and double-layered membranous structure called the .... *omentum*.

56

3. In the midline behind the stomach and the peritoneal cavity are the duodenum and a glandular organ called the _pancreas_.

4. Connected with the intestine is a double-layered peritoneal structure called the .... _messentery_

5. At the lower ventral part of the abdominal cavity and below the greater peritoneal cavity is a hollow organ called by the 2-word name .... _urinary bladder_

6. The extension of the greater peritoneal cavity behind the reproductive organs has a French name of 3 words that describes its structure. This name is .... _cul-de-sac_.

VI. PRACTICAL APPLICATIONS

Study each discussion. Then print the appropriate word or phrase in the space provided.

Group A

While observing in an outpatient clinic a student noted the following cases.

1. Baby J experienced difficulty in breathing and copious discharge from his nose. A diagnosis of U.R.I. (upper respiratory infection) was made. The location of the membrane and the type of discharge indicated that the involved membrane was one of the .... _mucous (pleural.)_

2. Mrs. K complained of a swelling in the left groin. She had suffered previously from an infection of bone in the middle back; now it appeared that the infection had traveled along the fibrous covering of some of the back muscles. Such muscle coverings are called .... _muscle sheaths_

3. Mr. B was concerned about swelling and tenderness over his neck and upper back. His work involved the demolition of old buildings; he had become careless about personal cleanliness. Infection now involved the skin and connective tissue under it. The "sheet" that underlies the skin is called .... _superficial fascia_

4. Miss G complained of sharp pains in the chest and one side. Her disorder was diagnosed as inflammation of the membrane that forms a sac around each lung. This membrane is the .... _pleura_.

5. Mrs. J had suffered a painful bump on her ankle. The swelling involved the superficial tissues and the fibrous covering of the bone, or the .... _periosteum_

57

Group B

While working in an intensive care unit the nurse reported on the following cases.

1. Mr. B had suffered acute abdominal pain and other symptoms of appendicitis. Because of his delay in seeking treatment his appendix ruptured. He required exacting care following surgery because of a serious infection of the abdominal serosa. This disorder is called .... *peratinitis*.

2. Mrs. C had undergone extensive surgery because of deformities due to rheumatoid arthritis, an inflammatory disorder of the membranes lining the joint spaces. These lining membranes are known as .... *synovial*.

3. Mr. M had a history of repeated bacterial infections that involved the lining of organs of the reproductive system. Now, because of neglect, his urinary system was also affected. The continuous lining found in the reproductive and urinary systems is classified as .... *mucous membrane*.

4. Miss G experienced abdominal pains following longstanding infection of the pelvic organs. Connective tissue bands (adhesions) were found to extend throughout the peritoneal surface. The layer of peritoneum that is attached to the organs is called the .... *visceral*.

5. Miss G complained also of pain with certain motions. This was probably due to the pull of the adhesions on the nerve endings in the abdominal wall. The layer of serosa lining the wall is the .... *parietal*.

6. Student N suffered a mild concussion while playing football and it was feared that there might be damage to the brain coverings. These brain and spinal cord coverings are known as .... *meninges*.

VII. REFERENCES

Memmler, R. L., and Wood, D. L.: The Human Body in Health and Disease, ed. 4, pp. 53-57. Philadelphia, Lippincott, 1977.

Chaffee, E. E., and Greisheimer, E. M.: Basic Physiology and Anatomy, ed. 3, pp. 208-209. Philadelphia, Lippincott, 1974.

Crouch, J. E., and McClintic, J. R.: Human Anatomy and Physiology, pp. 59-67, 310, 389, 403, 404. New York, Wiley, 1971.

Greisheimer, E. M., and Wiedeman, M. P.: Physiology and Anatomy, ed. 9, pp. 29-31. Philadelphia, Lippincott, 1972.

CHAPTER SIX

The Blood

I. OVERVIEW

Blood is an important *indicator* of a person's *health*; some understanding of its constituents and their *functions* is needed by all those engaged in health occupations. Blood has the 2 functions of *transporting*, by bringing needed substances such as food materials and oxygen to all the body tissues, and carrying off waste products, and *combating infection*, by defending the body against harmful organisms and maintaining its disease immunity.

The *plasma* of the blood consists of water, protein, carbohydrates, lipids, mineral salts and some other substances which are needed for normal body function. The *formed elements* of the blood, called the *corpuscles*, are composed of the *erythrocytes* (the red cells, which carry oxygen to the tissues by means of their *hemoglobin*), the *leukocytes* (the white cells, which defend the body by engulfing harmful pathogens) and the *platelets* (the thrombocytes, which play an essential role in blood clotting). The corpuscles are mainly formed in the red bone marrow.

One reason for studying the characteristics of blood is to be able to recognize abnormal findings, because such blood findings indicate that certain diseases may be present. Anemias, neoplastic diseases of blood and hemorrhagic disorders are all associated with blood abnormalities.

A second reason is concerned with blood transfusions, which often become necessary in the treatment of various diseases. A procedure known as crossmatching is done to assure that the blood of the patient and that of the donor are compatible.

By adding sodium citrate to blood to prevent clotting, blood can be stored for a number of days to be used in times of emergency. Here, too, the blood must be typed and crossmatched before being transfused.

The presence or absence of the Rh factor, a red cell protein, is also determined by testing—another reason for our need to understand the characteristics of blood. If blood containing the Rh factor (Rh positive) is given to a person whose blood lacks that factor (Rh negative), the recipient may become *sensitized* to the protein; his blood will produce *antibodies* to counteract the foreign substance.

Blood group studies may also provide useful information about paternity in some cases.

In order to make the kinds of determinations mentioned, numerous *blood studies* have been devised. Large, modern laboratories are equipped with *automatic counters*, which rapidly and accurately "count" blood cells; and with *automatic analyzers*, which measure enzymes, electrolytes and other constituents of blood serum. The *blood smear* is examined for the presence of parasites or abnormal blood cells. It also measures the approximate percentages of the various types of white cells, which is significant in certain disorders.

II. TOPICS FOR REVIEW

1. purposes of blood
2. blood plasma and its functions
   a. proteins
   b. carbohydrates
   c. lipids
   d. mineral salts
3. the formed elements and their functions
   a. erythrocytes
      (1) structure and function
      (2) purpose of hemoglobin
   b. leukocytes
      structure and function
   c. platelets (thrombocytes)
      origin and function
4. formation of corpuscles
5. blood typing and blood transfusion
   a. blood groups
   b. Rh factor
   c. determining fatherhood
   d. use of blood bank
   e. conditions requiring transfusion
6. blood derivatives
7. disorders of blood
   a. main groups
   b. characteristics
8. blood studies
   a. blood-counting machines
   b. hemocytometer
   c. blood slide
   d. blood chemistry machines
   e. platelet count
   f. clotting time
   g. bleeding time
   h. sternal puncture

III. MATCHING EXERCISES

Matching only within each group, print the answer in the space provided.

Group A

bone marrow	thrombocytes	carbon dioxide
oxygen	plasma	erythrocytes
hemoglobin	leukocytes	nucleus

1. The liquid part of the blood is known as .... *plasma*.

2. The red blood cells are called .... *erythrocytes*

3. There are several types of white blood cells or .... *leukocytes*.

60

4. Elements that have to do with clotting include platelets, or _thrombocytes_

5. An important gas that is transported by the blood from the lungs to all parts of the body is .... _Oxygen_.

6. The blood carries a waste product to the lungs and this gas is known as .... _carbon dioxide_

7. An important ingredient of red blood cells is a compound called .... _hemogloben_.

8. A connective tissue present in bone is the site of formation of most blood cells. The name of this tissue is ...... _bone marrow_

9. The mature red blood cell differs from other body cells in that it lacks a .... _nucleus_

Group B

serum	fibrinogen	hemolysis
type O	hemoglobin	pathogens
megakaryocytes	type AB	agglutination

1. Oxygen, needed by all the tissues, is transported by the blood constituent .... _hemoglobin_.

2. The platelets are fragments of large cells known as ...... _megakaryocytes_

3. As platelets disintegrate they release a chemical that reacts with a plasma protein .... _fibrinogen_.

4. The process whereby cells become clumped is known as _hemolysis_.

5. The watery fluid that remains after a clot is removed is known as .... _serum_.

6. In blood transfusion a dangerous condition that occurs when donor cells are dissolved or go into solution is .... _agglutination_

7. Blood that is not clumped by either anti-A or anti-B serum belongs to the group called .... _type O_.

8. If the cells are clumped by both the anti-A and anti-B serums the blood belongs to .... _type AB_.

9. The appearance of pus at a body site indicates that the leukocytes are actively involved in the destruction of .... _pathogens_ ~~agglutination~~

Group C

purpura	malaria	transfusion
leukemia	anemia	hemorrhage
centrifuge	bone marrow	vitamin B_{12}

1. When the blood is lacking in the normal number of red blood cells or in overall quantity, the condition is referred to as .... *anemia*.

2. An abnormal increase in the number of immature white cells is seen in the neoplastic disease called .... *leukemia*.

3. Profuse bleeding is usually referred to as .... *hemorrage*.

4. The transfer of whole blood from one person to another is called a .... *transfusion*.

5. Separation of blood plasma from the formed elements of blood is accomplished by use of the .... *centrifuge*.

6. In the condition known as pernicious anemia the body is unable to absorb a substance essential for the formation of red blood cells. This is .... *vitamin B12*.

7. Excessive destruction of red blood cells may cause anemia, a symptom often found in the disease .... *malaria*.

8. In the condition known as aplastic anemia there is failure of the blood-forming organ, namely the .... *bone marrow*.

9. A disorder in which there are hemorrhages into the skin and mucous membranes is .... *purpura*.

Group D

hyperglycemia	leukocytosis	hemoglobinometer
leukopenia	hemocytometer	4.5 to 5.5 million
5,000 to 9,000	hematocrit	blood chemistry

1. An apparatus made of several parts and used for counting blood cells is called a .... _____.

2. Normally, the number of red blood cells per cubic millimeter is .... *4.5 to 5.5 million*

3. Normally, the number of white blood cells per cubic millimeter is .... *5,000 to 9,000*.

4. Blood normally contains some sugar. When the amount is excessive, the condition is referred to as .... *hyperglycemia*

5. The volume percentage of red blood cells in centrifuged whole blood is called the .... _____.

6. Determining the amounts of cholesterol, electrolytes, enzymes and glucose is a part of .... _____.

62

7. In most infections, as well as in various other types of illness, the white count may be excessive. This finding is referred to as .... _____.

'8. An abnormal reduction of the white blood count to below 5,000 is called .... *leukopenia*.

9. The apparatus that determines the amount of hemoglobin in the blood is known as the hemometer or .... _____.

IV. LABELING

For each of the following illustrations, print the name or names of each labeled part on the numbered lines.

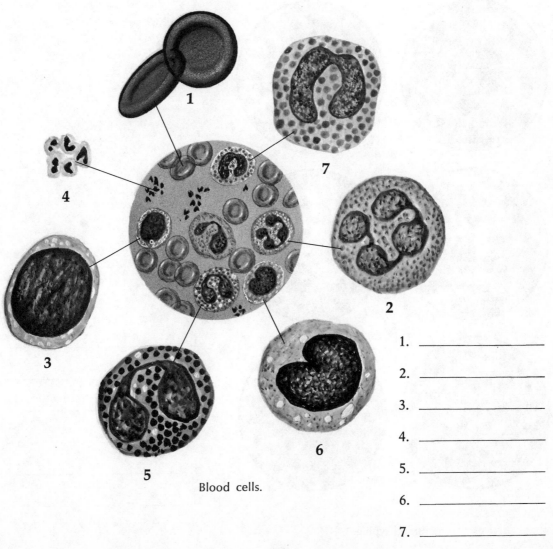

Blood cells.

1. _____
2. _____
3. _____
4. _____
5. _____
6. _____
7. _____

63

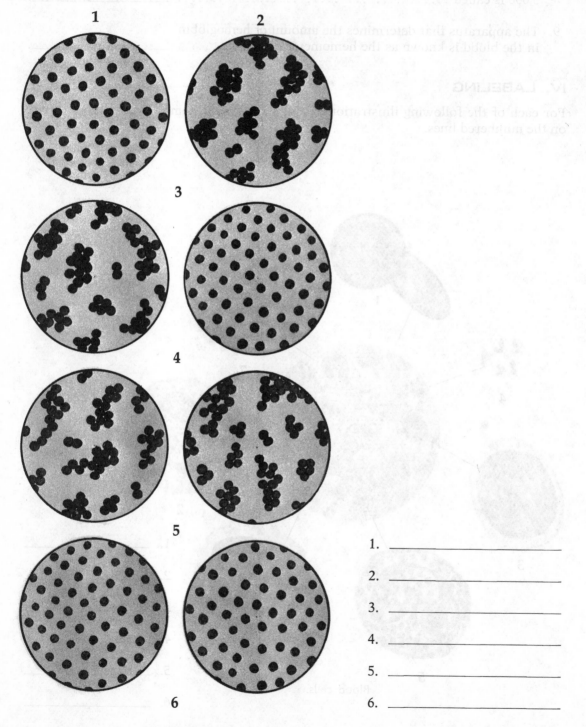

1

2

3

4

5

6

Blood typing.

1. _____

2. _____

3. _____

4. _____

5. _____

6. _____

1. _____
2. _____
3. _____
4. _____
5. _____
6. _____
7. _____

8. _____
9. _____
10. _____
11. _____
12. _____

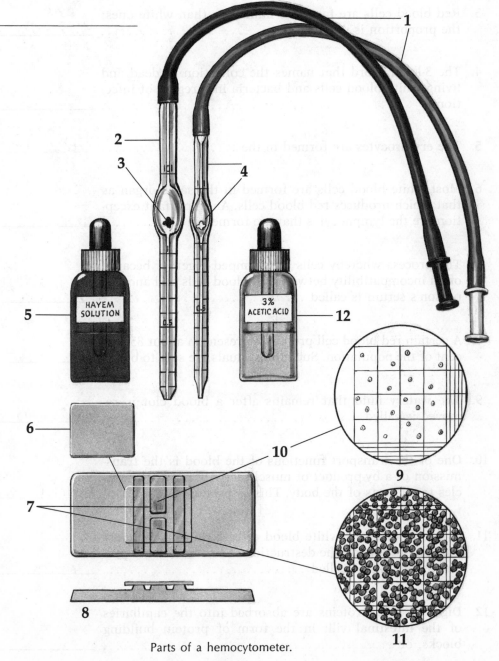

HAYEM
SOLUTION

3%
ACETIC ACID

Parts of a hemocytometer.

V. COMPLETION EXERCISE

Print the word or phrase that correctly completes the sentence.

1. The gas that is transported to all parts of the body by the blood and that is necessary for life is called . . . .   *oxygen* ~~carbon dioxide~~

2. One waste product of body metabolism is carried to the lungs to be exhaled. This gas is known as . . . .   *carbon dioxide*

3. Red blood cells are far more numerous than white ones; the proportion is, in fact, ...   *700-1*
   4.5-5 million
   red blood cells

4. The 3-letter word that names the collection of dead and living white blood cells and bacteria in a region of infection is . . . .   *pus* .

5. The erythrocytes are formed in the . . . .   *red blood cell*

6. Most white blood cells are formed in the same organ as that which produces red blood cells. An important exception are the lymphocytes that are formed in . . . . *red bone marrow*

7. The process whereby cells are clumped together because of an incompatibility between red blood cells and another person's serum is called . . . .   *hemolysis* .

8. A certain red blood cell protein is present in about 85 percent of the population. Such inidividuals are said to be . . . *Rh positive* .

9. The watery fluid that remains after a blood clot is removed is called . . . .   *serum* .

10. One of the transport functions of the blood is the transmission of a by-product of muscle activity from the muscles to all parts of the body. This by-product is . . . .   *heat* .

11. The leukocytes, or white blood cells, have as their most important function the destruction of certain disease producing organisms called . . . .   *pathogens* .

12. Digested food proteins are absorbed into the capillaries of the intestinal villi in the form of protein building blocks, or . . . .

66

VI. PRACTICAL APPLICATIONS

Study each discussion. Then print the appropriate word or phrase in the space provided.

Group A

1. Miss G sustained numerous deep gashes when she accidentally broke a glass shower door. One of the cuts bled copiously. In describing this type of bleeding the doctor used the word .... _Hemorrage_ .

2. While the physician attended to the wound the technician drew blood for typing and other studies. Miss G's blood was found to contain proteins that could be agglutinated. Her blood was classified as group .... _____.

3. Among the available donors were some whose blood was found to be free of agglutinable proteins. They were classified as having blood type .... _____.

4. Further testing of Miss G's blood revealed that it lacked the Rh factor. She was therefore said to be .... _RH neg._ .

5. If Miss G were to be given a transfusion of Rh positive blood, she might become sensitized to the Rh protein. In that event her blood would produce counteracting substances called .... _antibodies_ .

6. Mr. B had a history of peptic ulcer. On his admission he felt weak and was having severe abdominal pain. He was hospitalized and a series of tests was begun. One of these showed a reduction in the number of red blood cells and a decrease in the hemoglobin percentage. This condition is described by a word that means an insufficiency of blood, namely .... _____.

Group B

On the medical ward there were a number of patients who required extensive blood studies.

1. A boy 7 years of age had a history of frequent fevers and a tendency to bleed easily. Physical examination revealed enlarged lymph nodes. A blood smear revealed pronounced cell changes. The number of each kind of white cell was determined by counting the white cells in a sample of blood. This is called a .... _____.

67

2. Further study of this patient's blood smear revealed that numerous white blood cells were immature, and that the total number of white cells was tremendously increased. This disorder, a cancer of the blood, is .... _____

3. In order to obtain more information about the boy's illness, a sample of bone marrow was taken for study. A special needle was used to enter the breast bone. This procedure is known as .... *Sternal puncture*

4. Mrs. C's history included rapid weight loss, constant thirst and episodes of fainting. A blood test showed the presence of excessive sugar, or glucose. This condition is named .... _____

5. Mr. B, age 28, had a history of heart disease due to bacteria that caused dissolution (dissolving) of red blood cells. This type of disintegration is known as .... *Hemolytic*

6. Mr. K suffered from a viral infection of the liver. As a protective measure, his young son was given an injection of protein substance obtained from human plasma. This antibody, which prevents certain viral infections, has the name of .... *Fibrinogen*

7. Mr. Q complained of weakness and difficulty in walking. His red blood cell count was extremely low and abnormal red cells were found in the blood smear. His stomach hydrochloric acid was reduced. These signs are characteristic of a deficiency disease, a primary anemia called .... _____

8. A black child was brought in with complaints of swelling and pain in the fingers and toes. Blood studies revealed red cells that were somewhat crescent shaped and inflexible. These are characteristics of a disorder called .... _____

VII. REFERENCES

Memmler, R. L., and Wood, D. L.: The Human Body in Health and Disease, ed. 4, pp. 59-73. Philadelphia, Lippincott, 1977.

Chaffee, E. E., and Greisheimer, E. M.: Basic Physiology and Anatomy, ed. 3. Philadelphia, Lippincott, 1974.

Lenihan, J.: Human Engineering: The Body Re-examined, pp. 112-122, 135-138. New York, Braziller, 1975.

Nason, A., and Dehaan, R. L.: The Biological World, pp. 403-410. New York, Wiley, 1973.

Stonehouse, B.: The Way Your Body Works, pp. 38-39. London, Beasley, 1974.

CHAPTER SEVEN

Body Temperature and Its Regulation

I. OVERVIEW

Although heat is constantly being produced and lost during the course of the body's chemical activity, the body temperature is normally kept constant through its *homeostatic* mechanisms.

Homeostasis is the tendency of the body to maintain its stability despite the presence of forces that might tend to alter the situation. Through it the heart rate, blood pressure and the composition of the body fluids are also kept within the normal range.

Heat *production* is greatly increased during periods of increased muscular or glandular activity. Most heat *loss* occurs through the skin, with a smaller loss via the respiratory system and the urine and feces. The regulator responsible for keeping the temperature in a normal state regardless of heat production or heat loss is the *hypothalamus*, which transmits "messages" from the brain to the nerves so that the needed impulse is produced.

Although we usually think of the normal body temperature as being fixed at 98.6° F. (37° C.), it is more correct to speak of a *normal range*; the body temperature may vary with the time of day, the part of the body and the ingestion of food.

Abnormalities of body temperature are a valuable diagnostic tool. The presence of *fever*—an abnormally high body temperature—indicates infection most often, but may also indicate a toxic reaction, a brain injury and various other disorders. The opposite of fever is *hypothermia*—an exceedingly low body temperature—which most often comes about when the body is exposed to very low outside temperature and which can cause serious damage to the body tissues.

II. TOPICS FOR REVIEW

1. homeostasis, with examples
2. heat production within the body
3. heat loss
    a. conduction
    b. evaporation
    c. radiation
    d. convection

4. temperature regulation
5. normal temperature range
6. abnormal temperatures
    a. fever
        (1) causes
        (2) crisis
        (3) lysis
        (4) effect of phagocytosis
        (5) heat exhaustion
        (6) sunstroke (heat stroke)
    b. hypothermia
        (1) causes
        (2) effects

III. MATCHING EXERCISES

Matching only within each group, print the answer in the space provided.

Group A

| homeostasis | 36.2° to 37.6° C. | extremely low |
| oxygen | extremely high | muscles and glands |

1. Body heat is produced by the combination of food products with .... *oxygen* .

2. The largest amount of heat is produced in .... *muscles & glands*

3. In the condition of hypothermia, the body temperature is *extremely low*

4. The tendency of body processes to maintain a constant state is called .... *homeostasis*

5. The normal range of body temperature is .... *36.2° to 37.6° C*

6. In the condition of hyperthermia, the body temperature is *extremely high*

Group B

| skin | basal condition | blood |
| hypothalamus | tissues | subcutaneous fat |

1. The amount of heat produced by any organ depends partly on its activity and partly on its .... *tissues* .

2. Distribution of heat throughout the body is accomplished via the .... *blood* .

3. The body possesses several means of ridding itself of heat; the largest part of this loss occurs through the .... *skin* .

4. A natural insulator against cold is the .... *subcutaneous fat* .

5. The chief heat-regulating center in the brain is the ...... *hypothalamus*

6. The state of the body when it is at complete rest is known as the .... *basal condition*

Group C

hypothermia	sunstroke	infection
lysis	heat exhaustion	phagocytosis
crisis		

1. Excessive loss of salt may result in the condition of ..... *heat exaustion*

2. Failure of sweat glands to function when the body is exposed to high heat may result in .... *sunstroke* ~~heat exhaustion~~

3. Fever is most often due to .... ~~infection~~ *hypothermia*

4. Sometimes fever is beneficial because it steps up the process by which leukocytes destroy pathogens. This process is .... *phagocytosis*

5. A sudden drop in temperature at the end of a period of fever is referred to as .... *crisis*

6. Immersion foot is an example of the kind of injury that may be caused by prolonged .... *infection* ~~hypothermia~~

7. A gradual fall in temperature at the end of a period of fever is referred to as .... *lysis*

Group D

homeostasis	convection	radiation
evaporation	heat loss	heat gain
insulation	conduction	Celsius scale

1. Heat loss is accomplished in several ways. The transfer of heat from the body surface to the surrounding air is called .... *conduction*

2. Heat traveling from its source in the form of heat waves is called .... *radiation*

3. Freezing of water occurs at 0°, and the boiling point is at 100° in the .... *Celsius scale*

4. The amount of humidity has an effect on the rate of ..... *heat gain*

5. Muscular activity, as occurs during physical exercise, results in .... *heat loss*

71

6. The body has several ways of controlling its temperature. This regulation is one example of .... *homeostasis*.

7. When the layer of heated air next to the body is carried away and is replaced by cooler air the process is .... *convection*.

8. Clothing and subcutaneous fat represent different types of *insulation*.

9. Heat loss may occur during the conversion of a liquid or solid into a vapor, a process of .... *evaporation*.

IV. COMPLETION EXERCISE

Print the word or phrase that correctly completes the sentence.

1. While most heat loss occurs through the skin an appreciable amount is also lost in the urine, feces and via the ... _____.

2. The most important heat-regulating center is a section of the brain called the .... *hypothalamus*

3. Prolonged exposure to cold may result in an abnormally low temperature, a condition named the single word .... *hypothermia*

4. A gradual drop in the temperature of a fever is known as _*lysis*_.

5. Symptoms of rapid heat loss usually accompany a rapid drop in temperature which is known as .... *crisis*.

6. During basal conditions (when the body is at rest) the organ that is believed to produce about half the body heat is the .... *liver*.

7. One of the most important normal ways of increasing the production of body heat is by the activity of the many organs called .... _____.

8. A fever is usually preceded by a violent attack of shivering that is best described by the one 5-letter word ...... _____.

9. Because of the great increase in metabolism during a fever, it is important to give the patient a diet that can best be described by the 2 words .... *high calorie*

10. During a fever there may be considerable destruction of body tissues. Therefore, the diet should include foods that contain the nitrogenous compounds classified as .... _____.

11. Temperature control comes about in response to the heat brought to the brain by the blood as well as in response to impulses from the nerve endings in the skin called .... _____.

12. The formula for converting Fahrenheit temperatures to Celsius is .... $\frac{(C \times \frac{9}{5}) + 32}{(F - 32) \times \frac{5}{9}}$.

13. Practice changing Fahrenheit to Celsius. Show the figures for changing 50° and 70° F. .... ————————————.

14. Practice changing Celsius to Fahrenheit. Show the figures for 10° and 25° C. .... ————————————.

V. PRACTICAL APPLICATIONS

Study each discussion. Then print the appropriate word or phrase in the space provided.

A physician working in a desert area of southeastern California saw a variety of cases during the course of a day. The office nurse assisted him.

1. A 6-year-old male patient appeared apathetic and tired. His face was flushed and hot. On taking his temperature the nurse found it to be 105° F. The physician took the child's history and examined him, then instructed his mother to give the child cool sponge baths and administer the prescribed medication. The cool water sponging would aid in reducing the temperature. This is an example of .... ————————————.

2. Mr. C complained of painful areas over the tips of his toes. He informed the physician that he had worked in the mountains during the 3 winter months. The physician described these painful, itching red areas as .... ————————————.

3. A few men working on a construction project felt faint after working only half a day. They had been perspiring profusely but had not taken the salt tablets that were placed near the drinking faucet for their use. Such excessive salt loss results in .... ————————————.

4. Mr. K, age 69, had been working in his garden. The day was sunny and hot, but Mr. K neglected to protect his bald head by wearing a hat. He began to feel dizzy and faint. His wife noted that his face was very flushed, and his skin appeared dry. She put him to bed at once and called the physician, who informed her that these symptoms are typical of a disorder affecting the heat-regulating sweat glands and called .... ————————————.

5. Mrs. K was advised to apply an ice bag to her husband's head and to give him cool sponge baths in order to reduce ———————————.

73

VII. REFERENCES

Memmler, R. L., and Wood, D. L.: The Human Body in Health and Disease, ed. 4, pp. 73-80. Philadelphia, Lippincott, 1977.

Chaffee, E. E., and Greisheimer, E. M.: Basic Physiology and Anatomy, ed. 3, pp. 425-432. Philadelphia, Lippincott, 1974.

Ehrlich, P. R., Holm, R. W., and Soulé, M. E.: Introductory Biology, pp. 141-143, 245-246. New York, McGraw-Hill, 1973.

Schmidt-Nielson, K.: Animal Physiology, pp. 296-314. New York, Cambridge University Press, 1975.

CHAPTER EIGHT

The Skin in Health and Disease

I. OVERVIEW

Because of its various properties, the skin comprises an *enveloping membrane*, an *organ* and a *system*. A cross section of skin would reveal its layers of *epidermis* (the outermost layer), *dermis* (the true skin where the skin glands are mainly located) and the *subcutaneous fascia* (the under-layer).

The skin serves the essential functions of *protecting* deeper tissues against drying and against invasion by harmful organisms, *regulating* body temperature and *obtaining information* from the environment. It also *excretes* salt and water in the form of perspiration. The pigment *melanin* gives the skin its color; races that have been exposed to the tropical sun for thousands of years have highly pigmented skin.

The appearance of the skin is influenced by such factors as the quantity of blood circulating in the surface blood vessels and the hemoglobin concentration. Much can be learned about the condition of the skin by observing for the presence of *discoloration*, *injury* or *eruption*. Aging, exposure to sunlight and occupational activity also have a bearing on the condition and appearance of the skin. The skin is subject to numerous diseases, of which the most common are various types of *dermatitis*, such as *eczema* and *acne*.

Being the most visible aspect of the body, the skin is the object of much quackery, and vast sums of money are spent in efforts to beautify it; good general health is, however, the most important part of skin health and beauty.

II. TOPICS FOR REVIEW

1. the skin layers
   a. epidermis
   b. dermis
   c. subcutaneous layer (superficial fascia)
2. skin glands
   a. sudoriferous glands
   b. sebaceous glands
3. functions of the skin
4. general appearance of the skin
5. skin diseases

III. MATCHING EXERCISES

Matching only within each group, print the answer in the space provided. The same answer may be used more than once.

Group A

epidermis	sebaceous glands	melanin
corium	sudoriferous glands	tissue fluid
integument	subcutaneous layer	connective tissue

1. In its role as a system, the skin is called the .... *integument* .

2. Another term for dermis is .... ~~connective tissue~~ *corium*

3. Certain glands in the skin produce perspiration. These are the .... *sudoriferous*

4. Since the superficial fascia is the under-the-skin layer, it is also known as the .... *subcutaneous layer*

5. The oily secretion on skin and hair is produced by ...... *sebaceous glands*

6. Several layers of epithelial tissue form the outermost part of the skin, the .... *epidermis* .

7. Since the epidermis is lacking in blood vessels, nutritive substances reach the epidermal cells via .... *tissue fluid*

8. The framework of the dermis is composed of .... *connective tissue*

9. Skin color is due largely to the presence of pigment granules called .... *melanin*

Group B

trauma	receptors	dermis (or corium)
dilate	absorption	ciliary glands
pathogens	infection	

1. In its function of regulating body temperature the skin dissipates heat as the blood vessels enlarge or .... *dilate* .

2. Modified sweat glands are found in the eyelid edges. These are known as .... *ciliary glands*

3. In its role of protecting deeper tissues the skin prevents drying and invasion by .... *pathogens* .

4. The skin's function of obtaining information from the environment is due to the presence of a variety of sensory nerve endings. One general term for these is............. *receptors*

76

5. A framework of connective tissue and the presence of blood vessels and nerves characterize the .... _dermis)_ .

6. The general term for a wound or injury to the skin or other organs is .... _trauma_ .

7. The nerve endings of the skin are located mainly in the.. _receptors_ .

8. One of the functions of the skin that is actually very minimal is .... _absorption_ .

9. Following a wound or injury of the skin, pathogens may enter and cause an .... _infection_ .

Group C

intact	nerve endings	temperature regulator
melanin	ceruminous glands	perspiration and oil
more rapid	absorption	

1. Dilation of blood vessels brings more blood to the surface so that heat is dissipated into the air. This is one way in which the skin acts as a .... _temp. regulator_

2. The skin is an able defender against invasion by pathogens as long as it remains unbroken or .... _intact_ .

3. Exposure to sunlight causes an increase in the quantity of _perspiration_ + oil

4. Relatively, heat loss in the child compared with that in the adult is .... _more rapid_ .

5. A modification of the sweat glands is seen in the wax, or _ceruminous glands_

6. Medications are given by mouth or by injection more often than they are applied to the skin. This is because the skin has limited powers of .... _absorbtion_ .

7. Obtaining information about the environment is a function of the skin's .... _nerve endings_

8. The epidermal pores serve as outlets for .... _melanin_ .

Group D

crust	venereal herpes	urticaria
jaundice	vesicle	pustule
macule	papule	excoriation

1. Any flat discolored spot on the skin is called a .... _macule_ .

2. A yellowish skin discoloration may be due to the presence of bile in the blood. This condition is called............ *jaundice* .

3. The name given to a skin lesion produced by scratching is *excoriations*

4. A pimple that does not contain pus is called a........... *papules* .

5. A better term for scab is............................. *Crusts* .

6. A small sac that contains fluid is a blister or............ *Vesicles* .

7. Following the vesicular stage of chickenpox and smallpox pus appears in each lesion. This is a.................:....... *pustules* .

8. A viral skin disease that is characterized by many watery blisters is called .... *venereal herpes*

9. An allergic disorder in which there are itchy patches (hives) is called .... *urticaria* .

IV. LABELING

Print the name or names of each labeled part on the numbered lines.

1. _____ 9. _____

2. _____ 10. _____

3. _____ 11. _____

4. _____ 12. _____

5. _____ 13. _____

6. _____ 14. _____

7. _____ 15. _____

8. _____

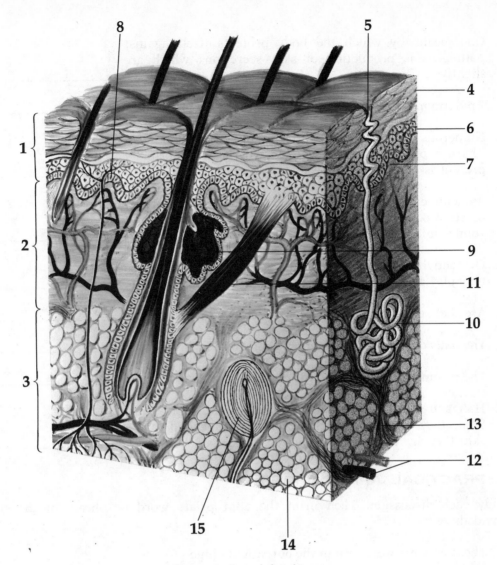

Cross section of the skin.

V. COMPLETION EXERCISE

Print the word or phrase that correctly completes the sentence.

1. The outer cells of the epidermis, which are constantly being shed, are designated the horny layer, or . . . . ————————————.

2. Many kinds of irritants and pathogens may cause skin inflammation. The general term for this disorder is . . . . ————————————.

3. A comedo is a plug of dirt and oily secretion in the pore of an oil gland. The common name is . . . . ————————————.

4. The pigment of the skin is . . . . ————————————.

5. Besides causing sunburn, prolonged excessive exposure to sunlight is thought to be one cause of . . . . ————————————.

6. One means by which the body protects itself against pathogens is production of skin secretions which are slightly .... _____.

7. Epidermophytosis, or athlete's foot, is due to infection by _____

8. Numerous factors including infection may cause baldness. Absence of hair from any areas where it is normally present is called .... _____.

9. An acute contagious skin disease caused by staphylococci or streptococci may be extremely serious in infants and young children. This disease is......................... _____.

10. Overactivity of the sebaceous glands during adolescence may play a part in the common skin disease............. _____.

11. Areas of redness represent a type of skin injury known as _____.

12. The nails originate from the outer part of the............. _____.

13. The ceruminous glands and the ciliary glands are modified _____.

14. Hair and nails are extensions from the outer epidermis. Since they are parts that are added or appended to the skin they are called cutaneous......................... _____.

VI. PRACTICAL APPLICATIONS

Study each discussion. Then print the appropriate word or phrase in the space provided.

These patients were seen in the outpatient clinic.

1. Mrs. A brought her 3 children to the clinic. The 9-month-old baby had redness of both cheeks, a symptom that the physician described as .... _____.

2. The physician also found several tiny blisters on the baby's cheeks, and made the notation on the chart that these eruptions were .... _____.

3. On both the baby's cheeks there were small pimplelike protrusions. The physician referred to them as........... _____.

4. The physician's diagnosis was that the baby had a common allergic disorder, ... _____.

5. Mrs. A's 6-year-old son had a number of blisters on his hands which contained pus. Microscopic examination revealed the presence of staphylococci. This contagious skin disease is called............................... _____.

80

6. The doctor ordered that the boy be kept home from school and that special care be followed to prevent infection in the 2 younger children. The middle child, age 4, seemed well. The physician was mainly concerned about the baby, since this disorder is highly contagious and may become so serious in a young baby as to be............. _____.

7. Mrs. A wondered why her own skin looked yellower than normal. Questioning indicated that she had become a food faddist and was eating carrots and other deeply colored vegetables to the exclusion of other foods. Mrs. A's condition is called .... _____.

8. L, age 15, came to the clinic with his father. The son's skin was marked by pimples and blackheads, and had a roughened appearance. This common disorder of the oil glands is found mainly in adolescents and is called....... _____.

9. L's father showed the physician several tumorlike masses on his scalp. These cysts, or sacs, which contain oil are referred to as .... _____.

10. Mr. K, age 38, was losing his hair. He mentioned that many of his male relatives had a similar problem. The medical term for baldness is.......................... _____.

11. Numerous vesicles and ulcers were found on Mrs. D's feet. This disorder is popularly called athlete's foot. A better term is .... _____.

12. Mrs. E's disorder was characterized by the presence of blisters along the course of certain nerves. This painful condition is commonly called shingles. The 2-word scientific name is .... _____.

13. Mr. M, a laborer, had neglected to give his skin proper care. Now numerous painful nodules were seen in the axillae. These nodules, due to bacteria entering the hair follicles, are called boils, or.......................... _____.

14. There was also a deep-seated infection of the subcutaneous fascia on Mr. M's lower back. These infected areas are known as .... _____.

15. Fair-skinned Mr. G had spent most of his 40 years in Arizona and southeastern California. He now noticed a firm nodule on the edge of the left ear pinna and said it was increasing in size. A biopsy revealed a malignancy called _____.

VII. REFERENCES

Memmler, R. L., and Wood, D. L.: The Human Body in Health and Disease, ed. 4, pp. 81-88. Philadelphia, Lippincott, 1977.

Chaffee, E. E., and Greisheimer, E. M.: Basic Physiology and Anatomy, ed. 3, pp. 43-47. Philadelphia, Lippincott, 1974.

Lenihan, J.: Human Engineering: The Body Re-examined, pp. 44-58. New York, Braziller, 1975.

Stonehouse, B.: The Way Your Body Works, pp. 20-21. London, Beasley, 1974.

Villee, C. H., and Delthier, V. G.: Biological Principles and Processes, pp. 590-591, 596. Philadelphia, Saunders, 1971.

CHAPTER NINE

Bones, Joints and Muscles

I. OVERVIEW

Bones, joints and muscles are so closely interrelated in function that they are collectively called the *musculoskeletal system*. The undergirding, or framework, of a building under construction might be likened to the musculoskeletal system; in both, the superstructure is wholly dependent on the framework.

To understand how the musculoskeletal system functions, you should first visualize the *skeleton*, with its 2 main divisions, the *axial* skeleton and the *appendicular* skeleton. The bones of the skeleton come together at the *joints*, which are also classified into 2 types, the *freely movable* joints and the *immovable* or *fixed* joints. Encasing bones and joints are the skeletal muscles. In its totality, the musculoskeletal system *protects*, *supports* and furnishes *motive* power.

II. TOPICS FOR REVIEW

1. bones
   a. structure and function
   b. divisions of the skeleton
      (1) the head
      (2) the trunk
      (3) the extremities
   c. bone disorders
2. joints
   a. types
   b. structure and function
   c. joint disorders
3. skeletal muscles
   a. characteristics
   b. principal skeletal muscles
      (1) the head
      (2) the neck
      (3) the chest and back
      (4) the abdomen
      (5) the lower extremities

c. muscle disorders
    4. practical applications
        a. importance and values of exercise
        b. levers and body mechanics

III. MATCHING EXERCISES

Matching only within each group, print the answer in the space provided.

Group A

cartilage	red marrow	yellow marrow
periosteum	calcium salts	appendicular skeleton
axial skeleton	endosteum	vertebral column

1. The bony framework of the head and trunk forms the.... _axial skeleton_

2. Production of blood cells is carried on mainly in the...... _red marrow_.

3. The combination of bones that form the framework for the extremities is called the.......................... _appendicular skeleton_

4. The fat found inside the central cavities of long bones is _yellow marrow_

5. The tough connective tissue membrane that covers bones is .... _periosteum_.

6. The somewhat thinner membrane that lines the central cavity of long bones is................................ _endosteum_.

7. The pliability of the young child's bones is due to their relatively large proportion of......................... _cartilage_.

8. A primary curve in the infant and secondary curves that develop in childhood are characteristic of the............ _vertebral column_

9. The brittleness of the old person's bones is due to their relatively large proportion of....................... _calcium salts_.

Group B

cranium	occipital bone	parietal bones
ethmoid bone	sutures	sphenoid bone
temporal bones		

1. The delicate spongy bone located between the eyes is called the .... _ethmoid bones_

2. At the back of the skull, and including most of the base of the skull, is situated the........................... _occipital bone_

3. That part of the skull which encloses the brain is the..... _cranium_.

84

4. The bat-shaped bone that extends behind the eyes and also forms part of the base of the skull is the............ *sphenoid bone*

5. The paired bones that form the larger part of the upper and side walls of the cranium are the.................. *parietal bones*

6. The 2 bones that form the lower sides and part of the base of the central areas of the skull are.............. *temporal bones*

7. The cranial bones join at places called.................. *sutures*.

Group C

mandible   maxillae   zygomatic bone
hyoid   nasal bones   lacrimal bone

1. At the corner of each eye is a very small bone, the....... *lacrimal bone*

2. The only movable bone of the skull is the.............. *mandible*.

3. Lying just below the skull proper is a U-shaped bone called the .... *hyoid*.

4. The higher part of each cheek is formed by a bone called the .... *zygomatic bone*

5. The 2 bones of the upper jaw are the.................. *maxillae*.

6. The 2 slender bones that form much of the bridge of the nose are the .... *nasal bone*.

Group D

true ribs   cervical section   floating ribs
lumbar part   vertebral column   thoracic section
rib cage   coccygeal part   scoliosis

1. The framework of the trunk includes the rib cage and the *vertebral column*

2. The spinal column is divided into 5 regions; the first 7 vertebrae comprise the main framework of the neck. This is the .... *cervical section*

3. Just below the first 2 sections of the vertebral column are 5 bones that are somewhat larger than the first 19 vertebrae. These form the .... *lumbar part*

4. The second part of the vertebral column has a distinct outward curve. These 12 bones comprise the............. *thoracic section*

5. Lateral curvature of the vertebral column is a common abnormality. It is called............................ _scoliosis_ .

6. In the child the tail part of the vertebral column is made of 4 or 5 small bones that later fuse. This is the........ _coccygeal_ .

7. Protecting the heart and other organs as well as supporting the chest are functions of the surrounding framework called the .... _Rib cage_ .

8. The first 7 pairs of ribs are called the.................. _true ribs_ .

9. Among the false ribs, as they are called, are 2 pairs, the last 2, which are very short and do not extend to the front of the body. These are the......................... _floating ribs_ .

Group E

patella	tibia	radius
ulna	olecranon	sesamoid

1. The upper part of the ulna forming the point of the elbow is the .... _____.

2. The medial forearm bone is the...................... _____.

3. The kneecap is also called the........................ _patElla_ .

4. Of the 2 bones of the leg the larger is the............... _tibia_ .

5. The forearm bone on the thumbside is the.............. _radius_ .

6. The patella is the largest of a type of bone that is encased in connective tissue. It is described as................. _____.

Group F

impacted fracture	foramen magnum	comminuted fracture
kyphosis	osteitis deformans	anterior fontanel
greenstick fracture	foramina	compound fracture
osteomyelitis	spiral fracture	lordosis

1. The skull of the infant, being in its formative stage, has a number of soft spots. The largest of these is the........ _anterior fontanel_

2. An infection of bone caused by pus-producing bacteria is _Osteomyelitis_

3. An incomplete break in a bone is most likely to occur in children. It is referred to as............................ _greenstick fracture_

4. Excessive concavity of the lumbar curve is.............. _lordosis_ .

5. The broken ends of the bones are jammed into each other in an .... *impacted fracture*

6. An abnormally increased curvature of the thoracic spine is called .... *Kyphosis*.

7. When tissues are torn and the bone protrudes through the skin the person is said to have a...................... *Compound fracture*

8. The largest opening in the skull, containing the spinal cord and related parts, is the........................... *foraman magnum*

9. A break in which there is more than one fracture line and several fragments are present is called.................. *Comminuted fracture*

10. If the bone has been twisted apart there is a........... *Spiral fracture*

11. An abnormality of body chemistry involving calcium is characteristic of Paget's disease, or................... *osteitis deformans*

12. Openings or holes that extend into or through bones are called .... *foramina*.

Group G

costae   pectoral girdle   phalanges
greater trochanters   calcaneus   carpal bones
metacarpal bones   symphysis pubis   ilium
pelvic girdle   processes   levers

1. The 5 bones in the palm of each hand are the........... *metacarpal bones*

2. The largest of the tarsal bones is the heel bone or........ *calcaneus*.

3. The 14 small bones that form the framework of the fingers on each hand are the........................ *phalanges*.

4. In the pelvic girdle, the os coxae is divided into 3 areas. The upper wing-shaped part is the..................... *illium*.

5. The bones of the wrist are the........................ *carpal bones*

6. The clavicle and the scapula are contained in the........ _____.

7. The os coxae articulating with the sacrum comprise the.. _____.

8. The ribs are also designated the...................... _____.

9. The pubic parts of the 2 ossa coxae unite to form the joint called the .... *Symphysis pubis*

10. The bones and muscles together form a system of........ _____.

11. The large rounded projections located at the upper and lateral portions of the femur are the.............. *greater trochanters*

12. Among the numerous prominences that serve as regions for muscle attachments are those called............... *processes*.

Group H

articular cartilage	flexion	ligaments
tendon	epimysium	synovial membrane
voluntary muscle	aponeurosis	rotation
extension	contraction	adduction
abduction	exercise	body mechanics

1. The lubricating fluid inside a joint cavity is produced by the cavity lining, the................................. *synovial membrane*

2. Bones in the region of a joint are held together by connective tissue bands called.......................... *ligaments*.

3. The contacting surfaces of each joint are covered by a layer of gristle, the................................. *articular cartilage*

4. A bending motion that decreases the angle between 2 parts is .... *flexion*.

5. Movement away from the midline of the body is known as *adduction*.

6. Motion around a central axis is called................... *rotation*.

7. When stimulated by nerve impulses, the muscle fibers become shorter and thicker; this results in muscle....... _____.

8. The rate of muscle metabolism increases during......... _____.

9. Energy may be conserved by using proper.............. _____.

10. Muscle may be attached to bone by a cordlike structure called a .... _____.

11. Sometimes a muscle is attached to a bone by means of a sheetlike structure, an .... _____.

12. The reverse of flexion is.............................. _____.

13. Another term for skeletal muscle is.................... _____.

14. The scissors kick of swimming is an example of......... _____.

15. The connective tissue sheath enclosing an entire muscle is the .... _____.

Group I

| articulation | motion | rheumatoid arthritis |
| rickets | bones | diarthroses |

1. Protection of delicate structures such as the brain is a function of .... _Bones_ .

2. A deficiency of calcium and phosphorus is the main cause of .... _rickets_ .

3. Certain joints allow for changes of position and thereby provide for .... _motion_ .

4. A most crippling inflammatory disease of joints is.. _rheumatoid arthritis_

5. The region of union of 2 or more bones is called a joint or an .... _articulation_ .

6. The more freely movable articulations are.............. _diarthroses_

IV. LABELING

For each of the following illustrations, print the name or names of each labeled part on the numbered lines.

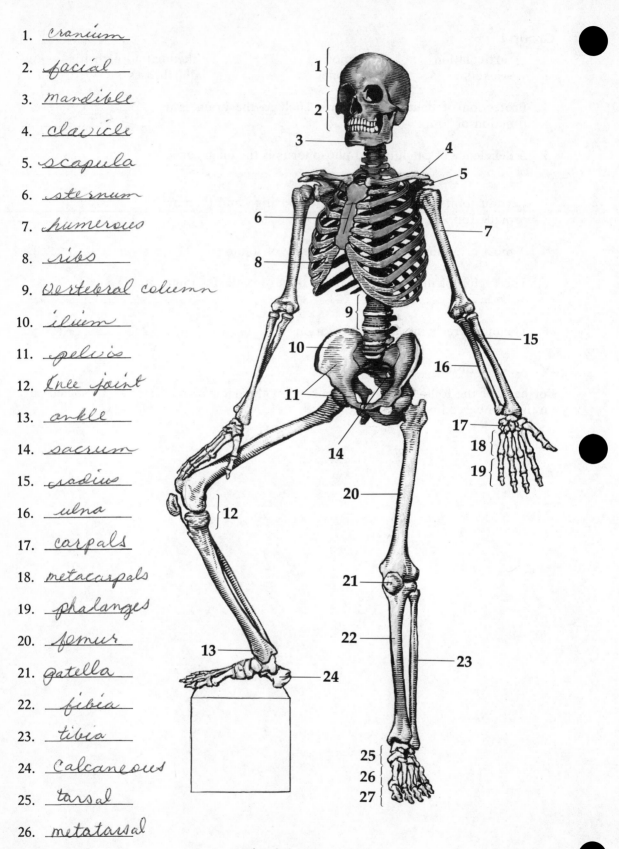

1. cranium
2. facial
3. mandible
4. clavicle
5. scapula
6. sternum
7. humerous
8. ribs
9. vertebral column
10. ilium
11. pelvis
12. knee joint
13. ankle
14. sacrum
15. radius
16. ulna
17. carpals
18. metacarpals
19. phalanges
20. femur
21. patella
22. fibia
23. tibia
24. calcaneous
25. tarsal
26. metatarsal
27. phalanges

The skeleton.

90

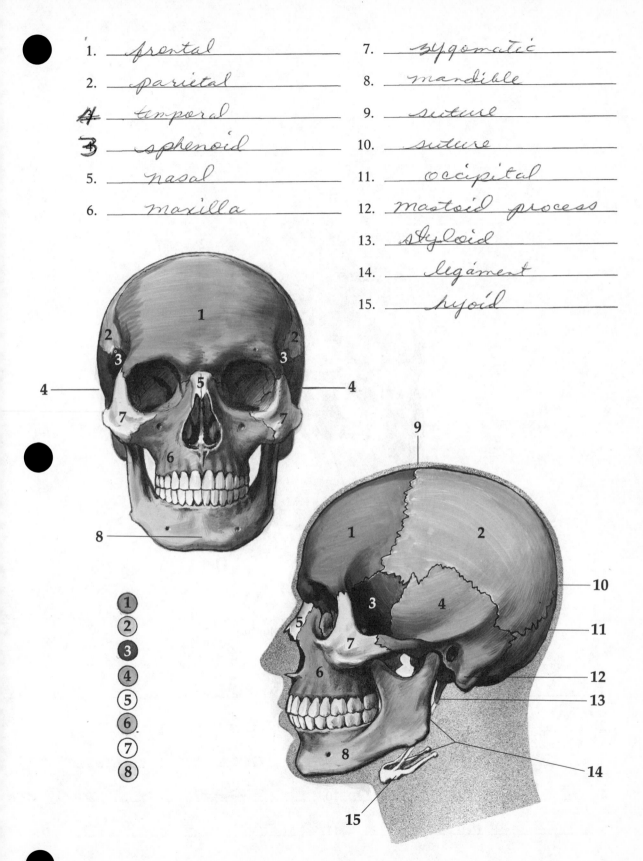

1. _frontal_
2. _parietal_
4. _temporal_
3. _sphenoid_
5. _nasal_
6. _maxilla_

7. _zygomatic_
8. _mandible_
9. _suture_
10. _suture_
11. _occipital_
12. _mastoid process_
13. _styloid_
14. _ligament_
15. _hyoid_

Skull from the front and from the left.

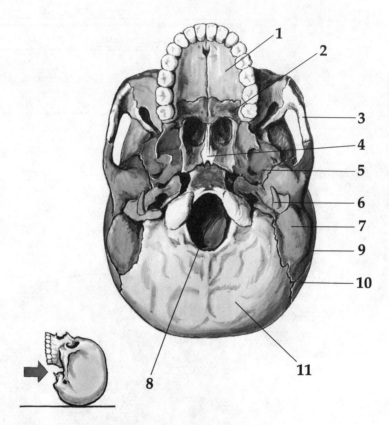

Skull from below, lower jaw removed.

1. _maxilla_
2. _palatine_
3. _zygomatic_
4. _vomer_
5. _sphenoid_
6. _styloid_

7. _mastoid_
8. _foramen magnum_
9. _temporal_
10. ~~occipital~~ _parietal_
11. _occipital_

92

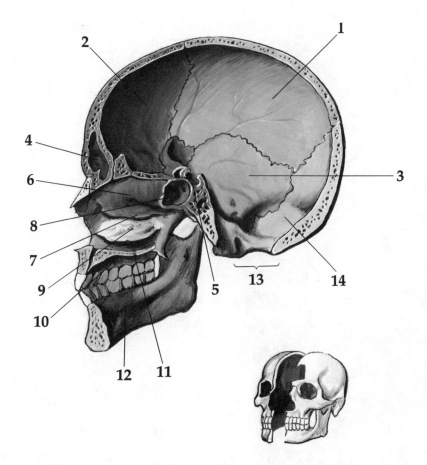

Skull, internal view.

1. _____   8. _____

2. _____   9. _____

3. _____   10. _____

4. _____   11. _____

5. _____   12. _____

6. _____   13. _____

7. _____   14. _____

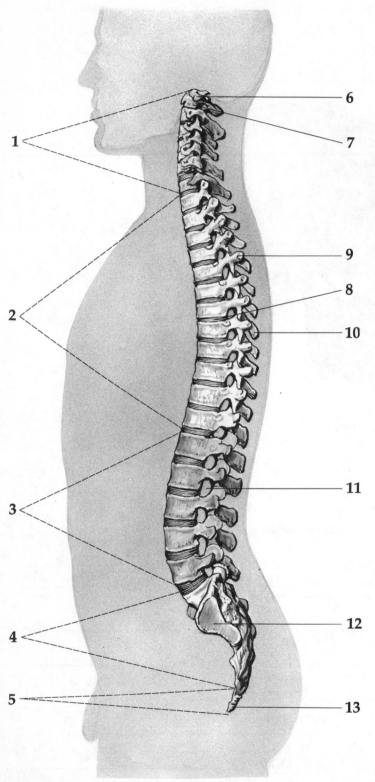

Vertebral column.

1. _cervical vertebrae_
2. _thoracic vertebrae_
3. _lumbar_
4. _sacral_
5. _coccygeal_
6. _atlas_
7. _axis_
8. _disk_
9. _transverse process_
10. _spinous process_
11. _foramen for spinal nerve_
12. _sacrum_
13. _coccyx_

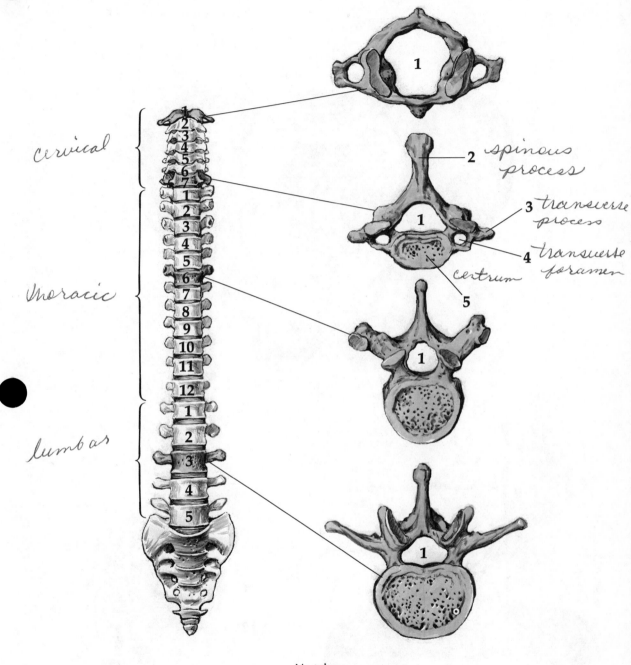

cervical

thoracic

lumbar

spinous process

transverse process

transverse foramen

centrum

Vertebrae.

1. _atlas_

2. _7ᵗʰ cervical Vertebra_

3. _____

4. _____

5. _____

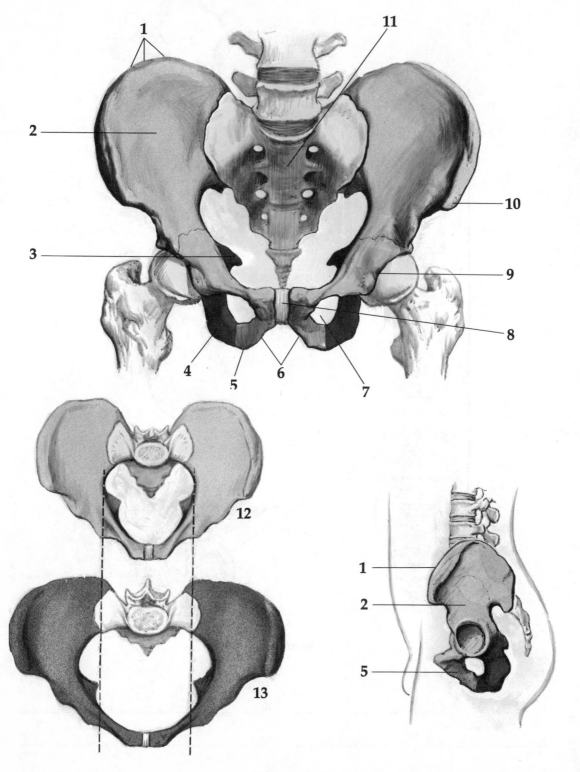

Pelvic girdle.

1. iliac crest
2. ilium
3. ischial spine
4. ischium
5. pubis
6. pubic arch
7. obturator foramen

8. symphysis pubis
9. socket for femur
10. iliac spine
11. sacrum
12. male pelvis
13. female pelvis

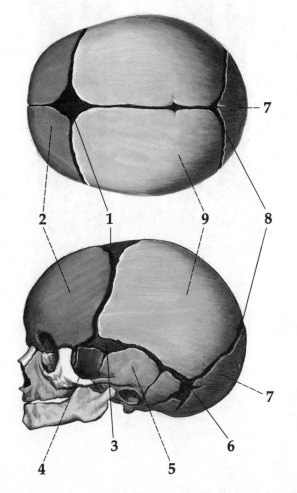

1. anterior fontanel
2. frontal
3. _____
4. _____
5. temporal
6. _____
7. occipital bone
8. _____
9. parietal

Infant skull, showing fontanels.

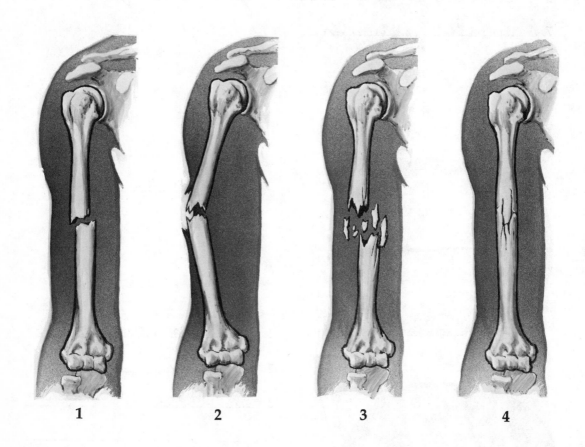

Types of fractures.

1. _____simple_____
2. _____compound_____
3. _____comminutes_____
4. _____greenstick_____

98

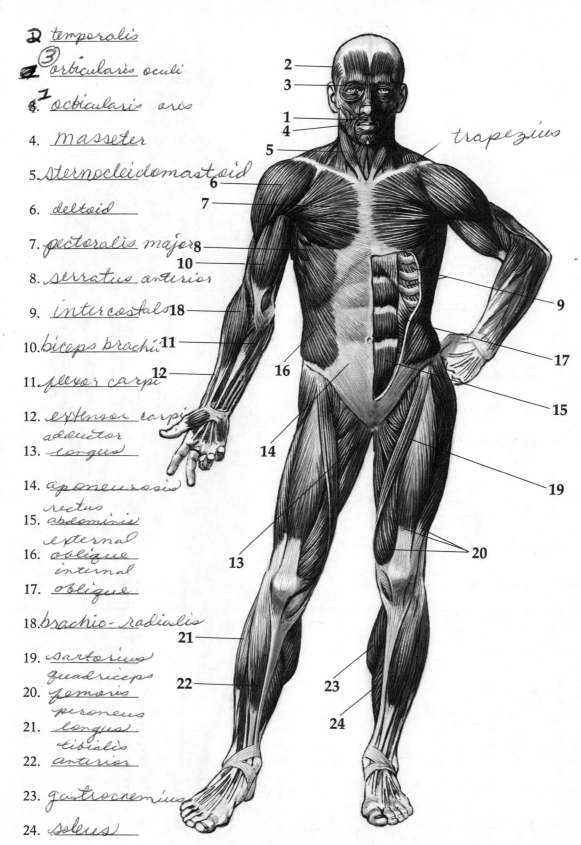

2. temporalis
3. orbicularis oculi
1. orbicularis oris
4. Masseter
5. sternocleidomastoid
6. deltoid
7. pectoralis major
8. serratus anterior
9. intercostals
10. biceps brachii
11. flexor carpi
12. extensor carpi
13. adductor longus
14. aponeurosis
15. rectus abdominis
16. external oblique
17. internal oblique
18. brachio-radialis
19. sartorius
20. quadriceps femoris
21. peroneus longus
22. tibialis anterior
23. gastrocnemius
24. soleus

trapezius

Principal muscles (anterior view).

99

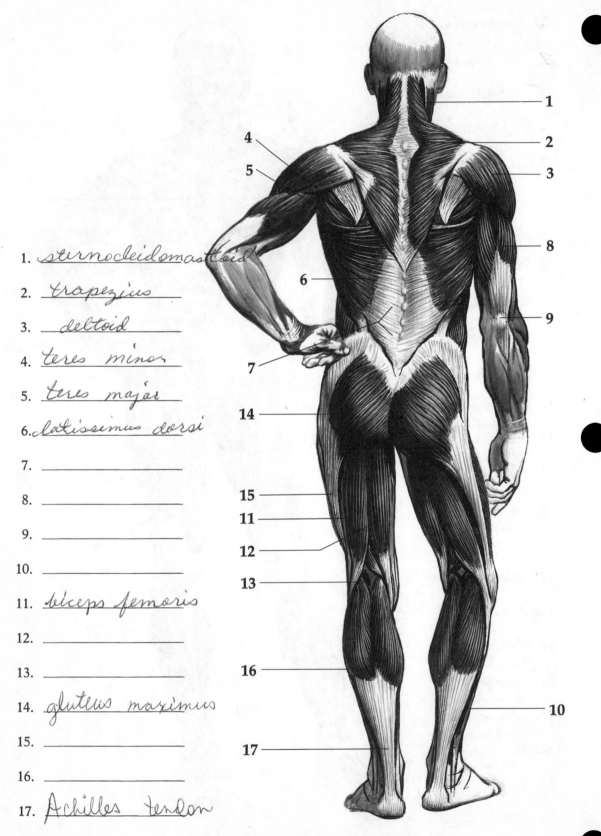

1. *sternocleidomastoid*

2. *trapezius*

3. *deltoid*

4. *teres minor*

5. *teres major*

6. *latissimus dorsi*

7. _____

8. _____

9. _____

10. _____

11. *biceps femoris*

12. _____

13. _____

14. *gluteus maximus*

15. _____

16. _____

17. *Achilles tendon*

Principal muscles (posterior view).

V. COMPLETION EXERCISE

Print the word or phrase that correctly completes the sentence.

1. Normally, muscles are in a partially contracted state, even though they are not in use at the time. This state of mild constant tension is called.......................... _____.

2. A movement is initiated by a muscle or set of muscles called the.......................... _____.

3. The movement of the prime mover is opposed by a muscle or set of muscles called the.......................... _____.

4. The muscle of the lips is the.......................... _____.

5. The flesh part of the cheek is formed by the............. _____.

6. There are 4 pairs of muscles of chewing. The muscles located at the angle of the jaw are called the........... _____.

7. A superficial muscle of the neck and upper back acts on the shoulder. This muscle is the.......................... _____.

8. Injury to the sternocleidomastoid may result in a disorder called wryneck, or .... _____.

9. The largest forearm extensor is the.......................... _____.

10. The muscular partition between the thoracic and abdominal cavities is the.......................... _____.

11. The large fleshy muscle of the buttocks which extends the hip is the .... _____.

12. The chief muscle of the calf is the toe dancer's muscle, named the .... _____.

13. The largest tendon in the body is the.......................... _____.

14. Bending the ankle so that the sole of the foot is facing outward, away from the body, is called................. _____.

15. The muscle attachment that is usually relatively fixed is called its .... _____.

16. Muscles contract and exert power on the more movable attachment, its .... _____.

17. Muscle fibers shorten and this contraction results in the actual movement of the muscle, its.................... _____.

18. The endings of the motor nerve fibers are called motor end plates or............................. _____.

VI. PRACTICAL APPLICATIONS

Study each discussion. Then print the appropriate word or phrase in the space provided.

Group A

A group of high school seniors were involved in a serious traffic accident on their way home from the prom.

1. There was a pronounced swelling of the upper right side of Mary's head. X-ray films showed a fracture of the largest skull bone, the................................. _____.

2. Mary also suffered an injury to one of the 2 large bones of the pelvic girdle. This bone articulates with the sacrum and is named the.................................... _____.

3. John suffered multiple injuries to his left lower extremity. Protruding through the skin was a splintered portion of the longest bone in the body, the...................... _____.

4. The muscle on the front of the thigh was involved in John's injury. This large 4-part muscle is the........... _____.

5. Susan thought her injuries were the least serious, so she walked several blocks to find help. Then she noticed that her right knee was not functioning normally. Examination revealed a fractured kneecap. Another name for the kneecap is .... _____.

6. Harry, the driver of the car, was forcibly thrown against the steering wheel. He suffered fractures of the sixth and seventh ribs, which are included among the............ _____.

7. Harry's chest was cut by flying glass. The largest anterior muscle in this area is the............................. _____.

Group B

Mr. B, age 58, was admitted to the general hospital because of acute pain and swelling of his right great toe. He also complained of a chronic backache. Mr. B underwent a complete physical examination.

1. Mr. B suffered from a disorder of metabolism in which uric acid accumulated in the blood and uric acid crystals were deposited in the joints of his right great toe. This disorder is called .... _____.

2. X-ray films showed involvement of the toe joints. The framework of the toes is made up of bones called the.... _____.

3. Spurs of bony material were found to be present at the edges of the vertebrae just above the sacrum, which is the part of the spinal column called the............... _____.

4. These spurs caused inflammation of the muscles of the back, a condition called............................. _____.

5. The muscle inflammation gave rise to muscle pain, or.... _____.

6. To add to these problems, there was inflammation of the connective tissues, a disorder known as................. _____.

Group C

Mrs. C, age 36, visited her doctor's office because of swelling and pain in the joints of her hands and fingers. Examination revealed the following:

1. Evidence of inflammation and overgrowth of the lining membrane of the joint cavities, a membrane that is called the .... _____.

2. Difficulty in moving the joints of the fingers due to damage to the normally smooth gristle on the joint surface. This layer is called the............................. _____.

3. That Mrs. C was probably suffering from the common disorder called .... _____.

VII. REFERENCES

Memmler, R. L., and Wood, D. L.: The Human Body in Health and Disease, ed. 4, pp. 91-115. Philadelphia, Lippincott, 1977.

Chaffee, E. E., and Greisheimer, E. M.: Basic Physiology and Anatomy, ed. 3, pp. 51-160. Philadelphia, Lippincott, 1974.

Lenihan, J., Human Engineering: The Body Re-examined, pp. 7-43. New York, Braziller, 1975.

Stonehouse, B.: The Way Your Body Works, pp. 22-27. London, Beasley, 1974.

CHAPTER TEN

The Brain, the Spinal Cord and the Nerves

I. OVERVIEW

The nervous system is the body's *coordinating system*, receiving, sorting out and responding to both internal and external stimuli. The nervous system as a whole is divided into the *central* nervous system, consisting of the brain and the spinal cord, and the *peripheral* nervous system, made up of the cranial and the spinal nerves. The central nervous system includes the *speech centers*, which are essential to one's ability to hear, see, speak and write, as well as certain parts that control such vital functions as respiration, heart rate and body balance. The peripheral nervous system controls both the *general* and the *special* sense impulses. Certain nerves and centers in this system are grouped together as the *autonomic* nervous system, because they control activities that go on more or less automatically, regulating the actions of glands, smooth muscle and the heart.

The whole of the central nervous system functions via billions of *neurons*, the structural units of the nervous system, each of which is composed of a *cell body* and *nerve fibers* that carry impulses to and away from the *cell body*.

II. TOPICS FOR REVIEW

1. structure and function of nervous system as a whole
   a. the nerve cell
   b. the nerve
2. divisions
   a. central nervous system
   b. peripheral nervous system
   c. autonomic nervous system
3. central nervous system
   a. the brain
      (1) main parts

105

        (2) cerebral hemispheres
        (3) cerebral cortex
        (4) speech
        (5) interbrain, midbrain, cerebellum, pons, medulla oblongata
        (6) ventricles
        (7) brain waves
    b. spinal cord
        (1) location
        (2) structure
        (3) function
        (4) spinal puncture
    c. coverings of brain and spinal cord
    d. cerebrospinal fluid
  4. disorders of the brain and cord
    a. stroke, cerebral palsy, epilepsy
    b. infections: meningitis, encephalitis
    c. tumors, multiple sclerosis
  5. peripheral nervous system
    a. cranial nerves
    b. spinal nerves
  6. autonomic nervous system
    a. parts
    b. functions

III. MATCHING EXERCISES

Matching only within each group, print the answer in the space provided.

Group A

brain and spinal cord	cerebral hemispheres
autonomic nervous system	nerve
nerve fibers	brain stem
stimuli	neuron
peripheral nervous system	coordinator

1. In relation to the parts and organs of the body, the
   nervous system functions as........................... _____.

2. The internal and external changes that affect the nervous
   system are called .... *stimuli*.

3. For study purposes, the entire nervous system has been
   divided into 2 large systems. One of these, the central
   nervous system, is composed of the .... *brain stem*.

4. The other large nervous system is the.................. *Peripheral N.S*

5. The sympathetic and parasympathetic nervous systems
   are the 2 functionally opposing parts of the............ *autonomic nervous system*

6. The neuron is composed of a cell body with the addition of threadlike cytoplasmic projections, the .... _____.

7. The cerebrum is the largest part of the brain. It is divided into right and left parts called the............. *cerebral hemispheres*

8. The midbrain, pons and medulla oblongata form the..... _____.

9. The basic nerve cell, including the cell body and its projections, is the .... *neuron* .

10. The peripheral nerves that regulate activities going on more or less automatically are grouped together as the... _____.

11. Impulses are conducted from one place to another by the bundle of nerve fibers, the........................... *nerve* .

Group B

sulci	mixed nerves	afferent nerves
receptor	convolutions	efferent nerves
cerebral cortex	lateral ventricles	synapse

1. The place where the stimulus is received is called the end organ or .... _____.

2. The region or junction at which an impulse is transmitted from one neuron to another is the.............. _____.

3. Impulses must be carried to and away from the brain and spinal cord. Those that conduct impulses to the brain and cord are grouped together as the.................. _____.

4. The nerve fibers that carry impulses away from the brain and cord to muscles and glands form the.............. _____.

5. All thought, association and judgment take place in the.. _____.

6. Among the important landmarks in the cerebral hemispheres are several fissures, or........................ _____.

7. The fissures serve to separate the gray matter in folds forming elevated portions known as.................... _____.

8. Cerebrospinal fluid fills spaces within the hemispheres. These are the .... _____.

9. Combinations of afferent and efferent fibers form........ _____.

107

Group C

internal capsule   temporal lobe   thalamus
lobes   motor cortex   parietal lobe
occipital lobe   meninges   myelin

1. The cerebral cortex of each hemisphere is divided into regions each of which regulates certain types of functions. These areas are called.............................. _____.

2. In the disorder known as multiple sclerosis there is degeneration of the fatlike substance that covers many nerve fibers. This sheath is.............................. _____.

3. The three brain coverings are collectively known as the.. _____.

4. In each frontal lobe is an area that controls voluntary muscles. This is the.............................. _____.

5. Pain, touch and temperature are interpreted in the sensory area which is contained in the .... _____.

6. Impulses received by the ear are interpreted in the auditory center, which is located in the .... _____.

7. Messages from the retina are interpreted in the visual area of the .... _____.

8. The white matter of the cerebral cortex consists of collections of nerve fibers, one group of which is particularly vulnerable to injury. This area is the .... _____.

9. Two masses of gray matter which are located in the diencephalon act as relay centers monitoring sensory stimuli. These 2 masses constitute the .... _____.

Group D

ventricles   corpora quadrigemina   myelinated nerve
muscles of respiration   encephalogram (or   fibers
cerebrum   ventriculogram)   medulla oblongata
electroencephalograph   cerebellum   hypothalamus
blood pressure   diencephalon   neurilemma

1. By cutting into the central section of the brain, one can see the interbrain, or .... _____.

2. The vermis and the 2 lateral hemispheres at the sides form the .... _____.

3. The 2 cerebral hemispheres form much of the largest part of the brain, the .... _____.

4. The 4 fluid-filled spaces within the brain are called ...... _____.

5. The respiratory, cardiac and vasomotor centers are found in the .... _____.

6. The vasomotor center affects muscles in the blood vessel walls and thus influences .... _____.

7. To aid in the diagnosis of tumors and other brain disorders an x-ray picture is used. It is called an .... _____.

8. The measurable electric currents produced by the activity of the brain cells are recorded by the .... _____.

9. A thin sheath that covers peripheral nerve fibers, but not those in the brain and cord, plays a part in repair. This covering is called .... _____.

10. Body temperature, sleep, the heart beat and water balance are among the vital body functions regulated by the ..... _____.

11. The respiratory center exerts control over the .... _____.

12. The relay centers for eye and ear reflexes are located in the midbrain. They are the 4 .... _____.

13. The pons is white in color because it is made largely of _____.

Group E

epilepsy   cerebral palsy   cerebrovascular accident
aphasia   encephalitis   paraplegia

1. The rupture of a blood vessel, thrombosis or embolism that causes destruction of brain tissue may be referred to as a stroke, cerebral apoplexy or a .... _____.

2. Abnormalities of brain function without apparent changes in nerve tissues are characteristic of a chronic disorder called .... _____.

3. A congenital disorder characterized by muscle involvement ranging from weakness to paralysis is known as ... _____.

4. The general term referring to inflammation of the brain is _____.

5. A spinal cord injury in which there is loss of sensation and of motion in the lower part of the body may result in _____.

6. When referring to loss of the power of expression by speech or writing we use the term .... _____.

Group F

afferent nerves   nerve cell bodies   efferent nerves
receptor   effector   nerve fibers

1. The internal section of the spinal cord is composed of gray matter consisting of the .... _____.

2. Surrounding the gray part is a larger area made up of white matter, the .... _____.

3. The spinal cord has several essential functions. One of these is to conduct sensory impulses upward to the brain in tracts within the cord. These impulses are brought to the cord by .... _____.

4. The spinal cord also functions as a pathway for conducting motor impulses from the brain downward in descending tracts. These motor impulses leave the cord via ..... _____.

5. The reflex pathway begins with the part of a sensory neuron called a .... _____.

6. The sensory neuron conducts an impulse to a central neuron which then transfers it to a motor neuron. This typical reflex pathway terminates in a gland or a muscle with an _____.

Group G

dura mater   arachnoid membrane   pia mater
meningitis   choroid plexuses   subarachnoid space
arachnoid villi   hydrocephalus   gliomas

1. The innermost layer of the meninges, the delicate connective tissue membrane in which there are many blood vessels, is the .... _____.

2. The weblike middle meningeal layer is the .... _____.

3. The outermost meningeal layer, which is the thickest and toughest, is also made of connective tissue. It is the ..... _____.

4. Inflammation of the brain coverings due to diplococci or other pathogenic bacteria is called .... _____.

5. The majority of brain tumors are derived from the neuroglia and are called .... _____.

6. Normally, the cerebrospinal fluid helps protect the brain and spinal cord against shock. This fluid is formed inside the brain ventricles by the .... _____.

7. Normally, the fluid flows freely from ventricle to ventricle and finally out into the .... _____.

8. The fluid is returned to the blood in the venous sinuses through the projections called .... _____.

9. Any obstruction to the normal flow of cerebrospinal fluid may give rise to .... _____.

Group H

visual area	auditory speech	visual speech
written speech	center	center
center	sensory area	left cerebrum

1. Pain, touch, temperature, size and shape are interpreted in the parietal lobe, in a section called the .... _____.

2. The understanding of words takes place with the development of a temporal lobe area known as the .... _____.

3. The muscles in the right side of the body are controlled by the .... _____.

4. Messages from the retina are interpreted in the region of the occipital lobe known as the .... _____.

5. The ability to read with understanding comes with the development of the .... _____.

6. The ability to write words, which usually is a late phase in a person's total language comprehension, is a function of the .... _____.

Group I

ganglion	sensory cell ganglia	roots
cervical plexus	plexuses	brachial plexus

1. Each spinal nerve is attached to the spinal cord by combinations of nerve fibers called .... _____.

2. The small masses of nerve cell bodies attached to each dorsal root are the .... _____.

3. A collection of nerve cell bodies usually found outside the central nervous system is a .... _____.

111

4. A short distance away from the spinal cord each spinal nerve branches into 2 divisions; the branches of the larger division interlace to form .... _____.

5. The shoulder, the arm, the wrist and the hand are supplied by branches from the .... _____.

6. Motor impulses to the neck muscles are supplied by the . _____.

Group J

parasympathetic nervous system   oculomotor nerve
sympathetic nervous system   trigeminal nerve
hypoglossal nerve   acoustic nerve
olfactory nerve   optic nerve
vagus nerve   facial nerve

1. Recall the functions of the autonomic nervous system. The part that acts to prepare the body for emergency situations is the .... _____.

2. The part of the autonomic nervous system that aids the digestive process is the .... _____.

3. Impulses controlling tongue muscles are carried by the .. _____.

4. General sense impulses to the face and head are carried through the 3 branches of the .... _____.

5. Sense fibers for hearing are contained within the .... _____.

6. The muscles of facial expression are supplied by branches of the .... _____.

7. The nerve that carries smell impulses to the brain is the _____.

8. The contraction of most eye muscles is controlled by the _____.

9. The visual nerve is called the .... _____.

10. Most of the organs in the thoracic and abdominal cavities are supplied by the .... _____.

IV. LABELING

For each of the following illustrations, print the name or names of each labeled part on the numbered lines.

1. _____   5. _____   9. _____

2. _____   6. _____   10. _____

3. _____   7. _____   11. _____

4. _____   8. _____   12. _____

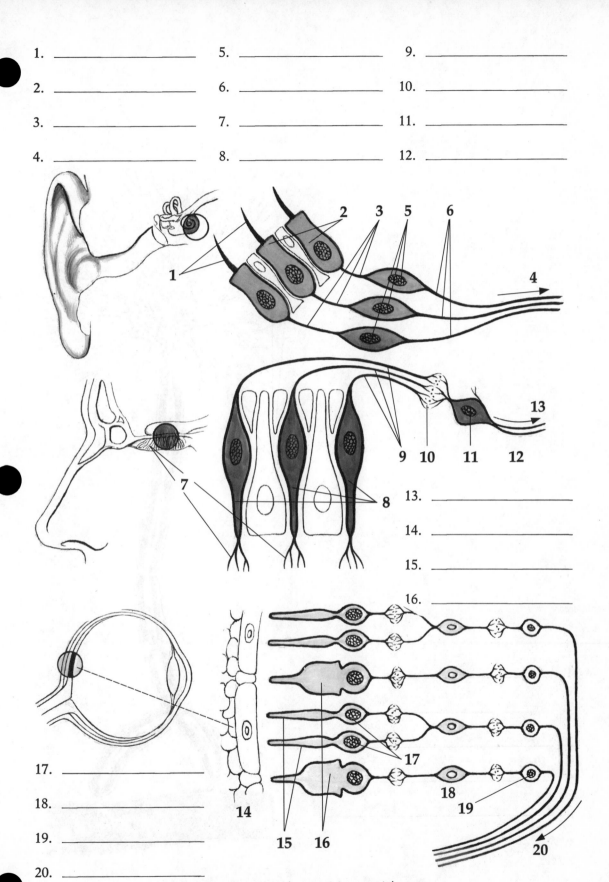

13. _____

14. _____

15. _____

16. _____

17. _____

18. _____

19. _____

20. _____

Diagram of neurons for receiving special senses.

113

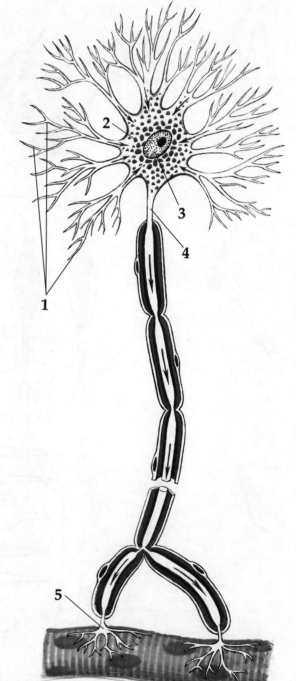

1. _____

2. _____

3. _____

4. _____

5. _____

Diagram of a motor neuron.

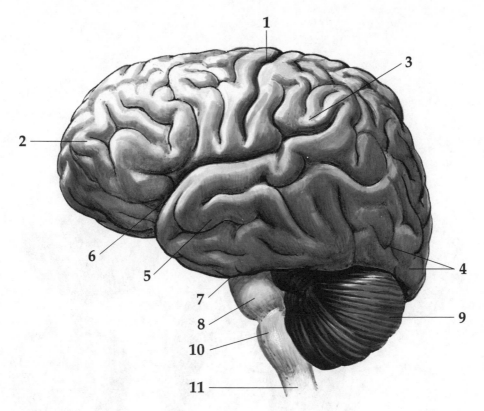

The external surface of the brain.

1. _____

2. _____

3. _____

4. _____

5. _____

6. _____

7. _____

8. _____

9. _____

10. _____

11. _____

115

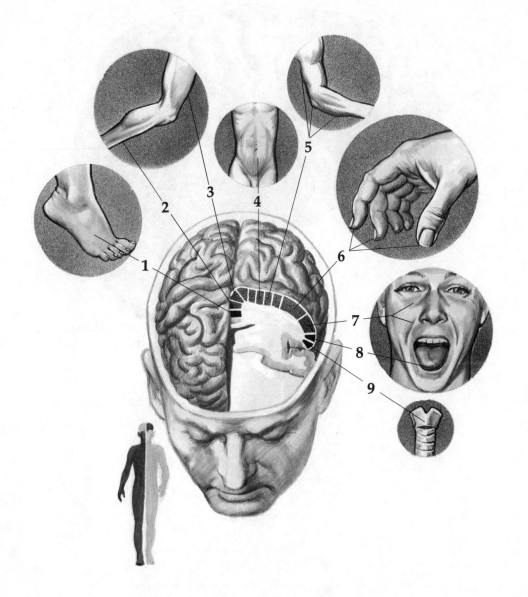

The motor area of the left cerebral hemisphere.

1. _____   6. _____

2. _____   7. _____

3. _____   8. _____

4. _____   9. _____

5. _____

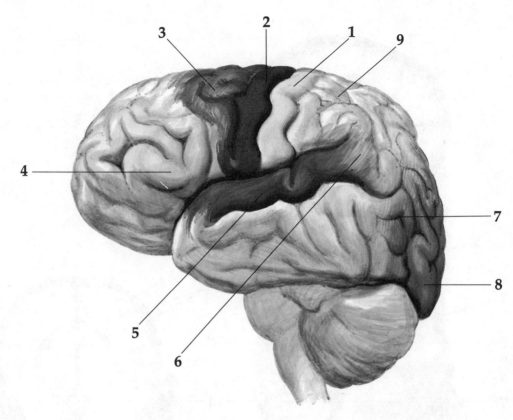

The functional areas of the cerebrum.

1. _____

2. _____

3. _____

4. _____

5. _____

6. _____

7. _____

8. _____

9. _____

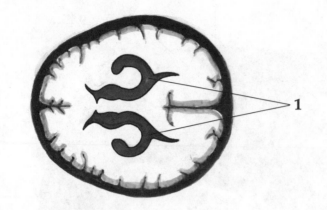

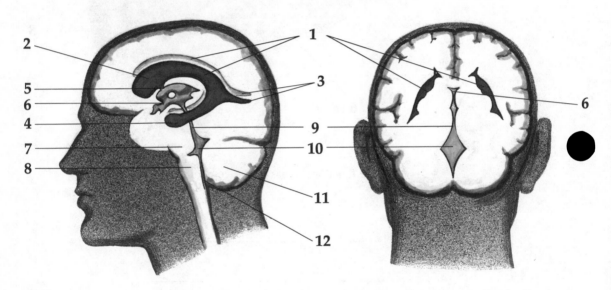

Brain ventricles.

1. _____   7. _____

2. _____   8. _____

3. _____   9. _____

4. _____   10. _____

5. _____   11. _____

6. _____   12. _____

Reflex arc and cross section of spinal cord.

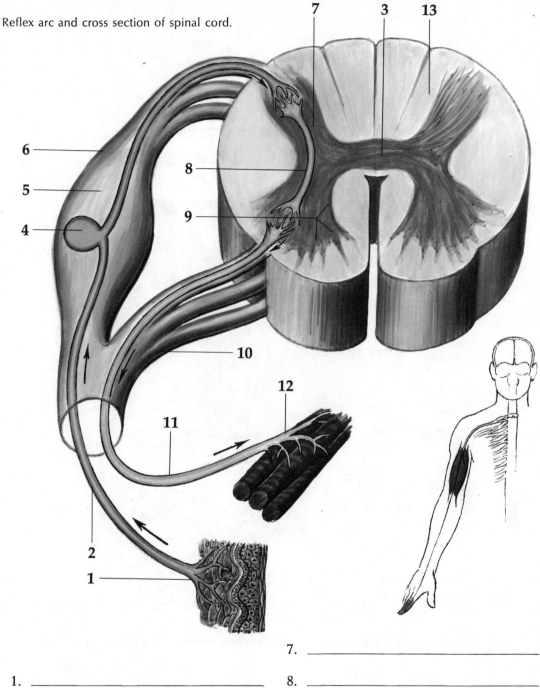

7. _____

1. _____ 8. _____

2. _____ 9. _____

3. _____ 10. _____

4. _____ 11. _____

5. _____ 12. _____

6. _____ 13. _____

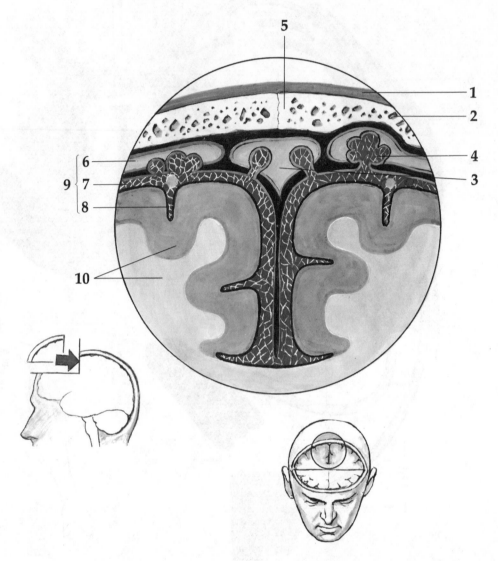

Frontal (coronal) section of top of head to show meninges and related parts.

1. _____   6. _____

2. _____   7. _____

3. _____   8. _____

4. _____   9. _____

5. _____   10. _____

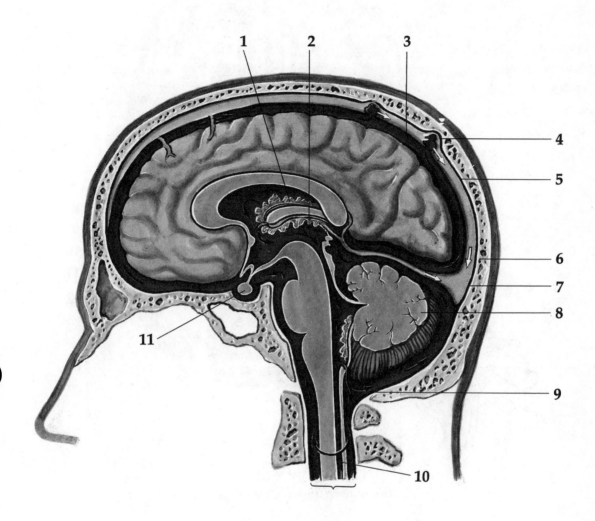

Flow of cerebrospinal fluid.

1. _____   7. _____

2. _____   8. _____

3. _____   9. _____

4. _____   10. _____

5. _____   11. _____

6. _____

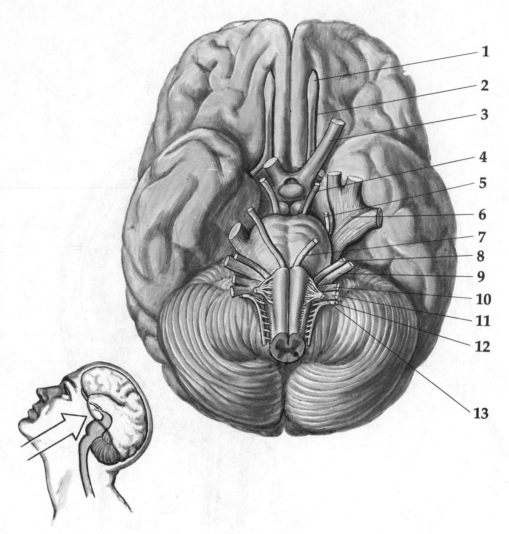

Base of brain, showing cranial nerves.

1. _____   8. _____

2. _____   9. _____

3. _____   10. _____

4. _____   11. _____

5. _____   12. _____

6. _____   13. _____

7. _____

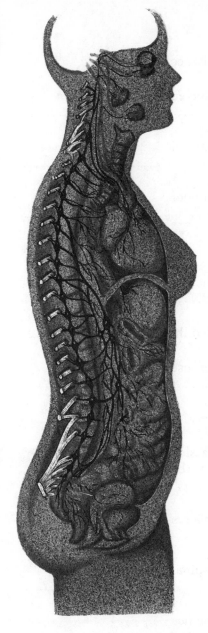

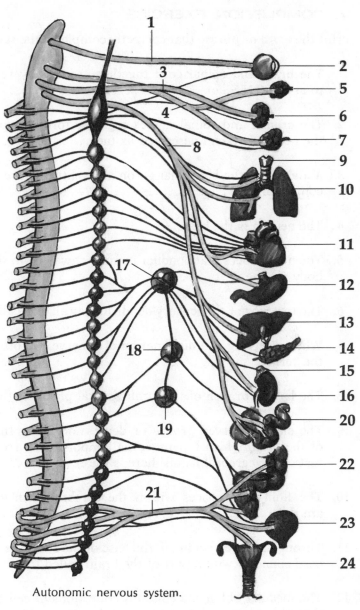

1 —

3
4

8

17

18

19

21

2 —
5 —
6 —
7 —
9 —
10 —
11 —
12 —
13 —
14 —
15 —
16 —
20 —
22 —
23 —
24 —

Autonomic nervous system.

1. _____	9. _____	17. _____
2. _____	10. _____	18. _____
3. _____	11. _____	19. _____
4. _____	12. _____	20. _____
5. _____	13. _____	21. _____
6. _____	14. _____	22. _____
7. _____	15. _____	23. _____
8. _____	16. _____	24. _____

V. COMPLETION EXERCISE

Print the word or phrase that correctly completes the sentence.

1. The brain and spinal cord together are usually referred to as the C.N.S., or .... _____.

2. The cranial and spinal nerves together form the part of the nervous system described as the .... _____.

3. Activities of the body that go on automatically are under control of the .... _____.

4. The neuron is the basic .... _____.

5. The nerve fibers that conduct impulses away from the cell body are the .... _____.

6. The receptor of the sensory dendrite may also be called the _____.

7. When referring to inflammation of a nerve, we often use the term .... _____.

8. The largest branch of the lumbrosacral plexus is the .... _____.

9. The slightly curved groove or depression along the side of the brain which separates the temporal lobe from the rest of the cerebral hemisphere is the .... _____.

10. The fluid-filled spaces within the cerebral hemispheres are the .... _____.

11. Through a large opening in the base of the skull the spinal cord connects with a part of the brain called the .... _____.

12. The medulla oblongata contains collections of cell bodies that have to do with important vital centers. These collections are known as centers, or .... _____.

VI. PRACTICAL APPLICATIONS

Study each discussion. Then print the appropriate word or phrase in the space provided.

1. Eight-year-old K was brought to the clinic because he had fallen during an epileptic seizure. There was bleeding from a scalp wound and some evidence of a subarachnoid hemorrhage. To aid in making a diagnosis the physician ordered a study of the boy's cerebrospinal fluid. This fluid is obtained by doing a .... _____.

2. Mrs. M's son found his mother lying unconscious on the floor. Mrs. M was 67 years old and had a history of high blood pressure. She was admitted to the intensive care unit. She was unable to speak or write, or to understand written or spoken language, so she was said to be suffering from .... ——————.

3. Mr. H, age 42, had been suffering for several weeks from persistent intractable headaches. An x-ray study of the brain was ordered. Such an x-ray is called an .... ——————.

4. Some of the fluid was removed from the ventricles in Mr. H's brain and replaced with air, as part of the diagnostic study. This fluid is the .... ——————.

5. As a result of the various studies done in Mr. H's case it was determined that a tumor was present in the left lateral ventricle. Surrounding the left ventricle is the .... ——————.

6. Miss S's symptoms included paralysis and various motor disturbances. The diagnosis of myelitis, or inflammation of the spinal cord was made. This nerve cord is located in a space called the .... ——————.

7. Mrs. J, age 60, was brought to the hospital with a diagnosis of stroke. In order to determine the location and extent of the hemorrhage, a test that uses high frequency sound impulses (echoencephalography) was done. It was found that there was damage on the left side of the brain which accounted for the paralysis of the opposite side of the body, a condition called .... ——————.

VII. REFERENCES

Memmler, R. L., and Wood, D. L.: The Human Body in Health and Disease, ed. 4, pp. 117-139. Philadelphia, Lippincott, 1977.
Chaffee, E. E., and Greisheimer, E. M.: Basic Physiology and Anatomy, ed. 3, pp. 161-229. Philadelphia, Lippincott, 1974.
Ehrlich, P. R., Holm, R. W., and Soulé, M. E.: Introductory Biology, pp. 424-437. New York, McGraw-Hill, 1973.
Lenihan, J.: Human Engineering: The Body Re-examined, pp. 163-183. New York, Braziller, 1975.
Stonehouse, B.: The Way Your Body Works, pp. 44-49. London, Beasley, 1974.
Villee, C. A., and Delthier, V. G.: Biological Principles and Processes, pp. 617-647. Philadelphia, Saunders, 1971.

CHAPTER ELEVEN

The Sensory System

I. OVERVIEW

Through the functioning of the *sensory receptors* we are made aware of all changes taking place both internally (within the body) and externally (outside the body). The *special* senses—so-called because the receptors are limited to a relatively small area of the body—include the visual sense, the hearing sense, the senses of taste and smell, those of hunger and appetite and the sense of thirst. The *general* senses are scattered throughout the body; they have to do with pressure, temperature, pain, touch and position.

II. TOPICS FOR REVIEW

1. the eye
   a. protective structures of eyeball
   b. coats of eyeball
   c. pathway of light rays
   d. sensory end organs (receptors)
   e. extrinsic eyeball muscles
   f. intrinsic eyeball muscles
   g. nerves
   h. lacrimal apparatus
   i. disorders, including blindness
2. the ear
   a. external ear
   b. middle ear
   c. inner ear
   d. disorders
3. other organs of special sense
   a. taste
   b. smell
   c. hunger and appetite
   d. thirst

4. general sense organs
    a. pressure
    b. temperature
    c. touch
    d. pain
    e. position

III. MATCHING EXERCISES

Matching only within each group, print the answer in the space provided. The same answer may be used more than once.

Group A

cornea   aqueous humor   transparent refracting parts
accommodation   rods and cones   choroid coat
retina   vitreous body   primary colors

1. The innermost coat of the eyeball, the nerve tissue layer, includes the end organs for the sense of vision. This structure is the .... _retina_ .

2. The pigmented middle tunic of the eyeball is the vascular _choroid coat_ .

3. Light rays pass through a series of transparent eye parts. The outermost of these is the .... _transparent refracting parts_

4. The watery fluid that fills much of the eyeball in front of the crystalline lens and also helps to maintain the slight curve in the cornea is the .... _aqueous humor_

5. The spherical shape of the eyeball is maintained by a jellylike material located behind the crystalline lens. This is the .... _vitreous body_

6. The receptors for the sense of vision are called the ...... _rods & cones_ .

7. There are 3 types of cones, each of which is sensitive to one of the .... _primary colors_

8. The elasticity of the lens enables it to become thicker and bend the light rays as necessary. This process is .... _accommodation_

9. The media of the eye may be described as the .... _____ .

10. Bulging forward slightly is the "window," or .... _cornea_ .

128

Group B

iris   pupil   ciliary body
media   sclera   receptors
optic disk   conjunctiva   diabetics

1. The opaque outermost layer of the eyeball is made of firm, tough connective tissue. This coat is the .... *sclera*.

2. The central opening in the iris contracts or dilates according to need. This opening is the .... *conjunctiva*.

3. Optic atrophy, cataracts and retinal disease are more common in persons who are .... *diabetics*.

4. The crystalline lens is one of the transparent refracting parts of the eye. Collectively they are called .... *media*.

5. The rods and cones of the retina are the visual end organs or .... *optic disk*.

6. The membrane that lines the eyelids is the .... _____.

7. The region of connection between the optic nerve and the eyeball is lacking in rods and cones and is commonly called the blind spot. Another term for this is .... _____.

8. The shape of the lens is altered by the muscle of the ..... *ciliary body*

9. The pupil is the central opening in the colored part of the eye, the .... *iris*.

Group C

fovea centralis   iris   intrinsic
sphincter   lacrimal gland   ophthalmia neonatorum
extrinsic   refraction   trachoma

1. The muscles that are attached to bones of the orbit and to the sclera are located outside the eyeball and are described as .... _____.

2. When a light is flashed in the eye the pupil is reduced in size due to the contraction of an iris muscle, the .... _____.

3. The amount of light entering the eye is controlled by the _____.

129

4. The process of bending which makes it possible for light from a large area to be focused on a small surface is known as .... _____ .

5. Tears serve an important protective function for the eye. They are produced by the .... *lacrimal gland* .

6. The clearest point of vision is a depressed area in the retina, the .... *fovea centralis*

7. An eye disease prevalent in poor and underdeveloped countries is characterized by the presence of granules on the lids. This serious disease is called .... _____ .

8. A suitable antiseptic prevents a serious eye infection of the newborn called .... _____ .

9. The muscles of the iris and ciliary body are located entirely within the eyeball and so are described as ...... _____ .

Group D

astigmatism	glaucoma	cataract
myopia	strabismus	lacrimation
opacity	hyperopia	crystalline lens

1. The eyes do not work together because the muscles do not coordinate in .... _____ .

2. Blurred vision and eyestrain is characteristic of the visual defect .... _____ .

3. The light rays are not bent sharply enough to focus on the retina, so that they cannot focus properly on close objects in farsightedness, or .... *hyperopia* .

4. The focal point is in front of the retina and distant objects appear blurred in nearsightedness, or .... *myopia, ~~hyperopia~~*

5. The lens loses its transparency and blindness ensues as a result of the formation of a .... _____ .

6. Continued high pressure of the aqueous humor may cause destruction of the optic nerve fibers. This cause of blindness is known as .... *glaucoma*

7. Removal of the cataract may restore useful vision. This involves removal of the .... _____ .

8. Injury or infection of the cornea may cause scar formation. Light rays cannot pass through the scar because there is now an area of .... _____.

9. The secretion of tears is called .... *lacrimation*

Group E

oval window	external auditory canal	eustachian tube
ossicles	endolymph	tympanic membrane
perilymph	pinna	mastoid air cells

1. Located at the end of the auditory canal is the eardrum, or _____.

2. The 3 small bones within the middle ear cavity are the ... _____.

3. The spaces within the temporal bone which connect with the middle ear cavity through an opening are called the .. _____.

4. Sound waves are conducted to the fluid of the internal ear by vibrations of the membrane that covers the .... _____.

5. Air is brought to the middle ear cavity by means of the auditory tube which is also called the .... _____.

6. The fluid of the inner ear contained within the bony labyrinth and surrounding the membranous labyrinth is called .... _____.

7. The fluid contained within the membranous labyrinth is called .... _____.

8. Sound waves enter the .... _____.

9. Another name for the projecting part, or auricle, of the ear is the .... _____.

Group F

optic nerve	vestibule	ophthalmic nerve
oculomotor nerve	cochlear duct	cochlear nerve

1. The organ of hearing is made up of receptors located in the .... _____.

2. The branch of the acoustic nerve that carries hearing impulses is the .... _____.

131

3. The entrance area that communicates with the cochlea and that is next to the oval window is the .... _____.

4. Visual impulses received by the rods and cones of the retina are carried to the brain by the .... _____.

5. Impulses of pain, touch and temperature are carried to the brain by a branch of the fifth cranial nerve, the ...... _____.

6. The largest cranial nerve carrying motor fibers to the eyeball muscles is the .... _____.

Group G

analgesic	ceruminous	olfactory
taste buds	polydipsia	pressure
adaptation		

1. The sense of taste involves 2 cranial nerves as well as receptors known as .... _____.

2. Excessive thirst, as may occur in certain illnesses including diabetes, is referred to as .... _____.

3. Among the general senses is that concerned with deep sensibility, commonly called the sense of .... _____.

4. In the case of many sensory receptors, including those for temperature, the receptors adjust themselves so that one does not feel the sensation so acutely if the original stimulus is continued. Such an adjustment to the environment is called .... _____.

5. Several methods, including use of drugs, are available for relief of pain. Aspirin is an example of the type of drug classified as .... _____.

6. The wax glands located in the external auditory canal are described as .... _____.

7. The pathway for impulses from smell receptors is the first cranial nerve, the .... _____.

IV. LABELING

For each of the following illustrations print the name or names of each labeled part on the numbered lines.

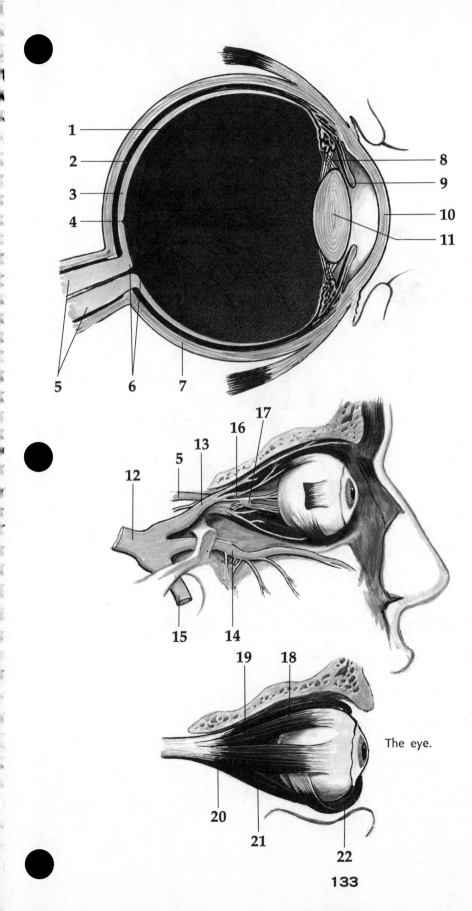

The eye.

1. _____
2. _____
3. _____
4. _____
5. _____
6. _____
7. _____
8. _____
9. _____
10. _____
11. _____
12. _____
13. _____
14. _____
15. _____
16. _____
17. _____
18. _____
19. _____
20. _____
21. _____
22. _____

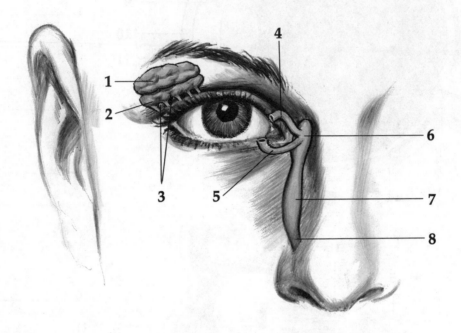

Lacrimal apparatus.

1. _____

2. _____

3. _____

4. _____

5. _____

6. _____

7. _____

8. _____

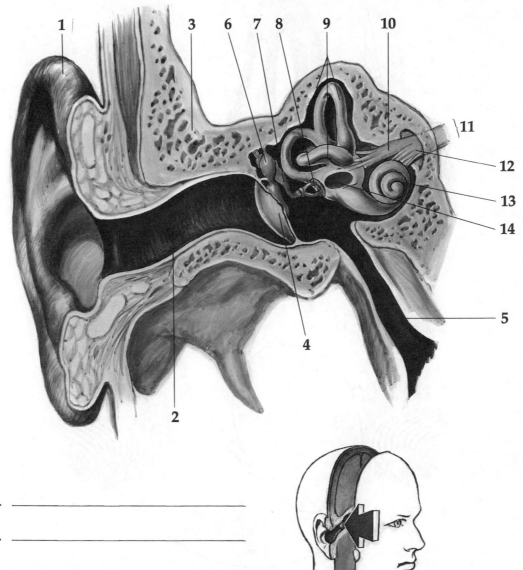

The ear.

1. _____

2. _____

3. _____

4. _____

5. _____

6. _____

7. _____

8. _____

9. _____

10. _____

11. _____

12. _____

13. _____

14. _____

135

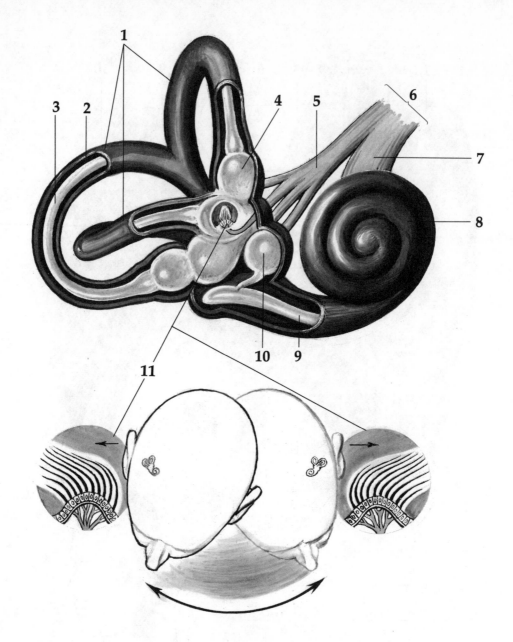

The internal ear.

1. _____   7. _____

2. _____   8. _____

3. _____   9. _____

4. _____   10. _____

5. _____   11. _____

6. _____

V. COMPLETION EXERCISE

Print the word or phrase that correctly completes the sentence.

1. The nerve fibers of the vestibular and cochlear nerves join to form the auditory nerve, more often called the . . . . _____.

2. The inner ear spaces contain fluids involved in the transmission of sound waves. The one that is inside the membranous cochlea and that stimulates the receptors is the _____.

3. The taste receptors of the tongue are located along the edges of small depressed areas, or . . . . _____.

4. The nerves involved in the sense of taste are the facial and the . . . . _____.

5. Pain that is felt in an outer part of the body such as the skin, yet that originates internally near the area where it is felt is called . . . . _____.

6. The very widely distributed free nerve endings are the receptors for the most important protective sense, namely that for . . . . _____.

7. The tactile corpuscles are the receptors for the sense of . . _____.

8. The nerve endings that relay impulses which aid in judging position and changes in location of parts with respect to each other are the . . . . _____.

9. The sense of position is partially governed by several structures in the internal ear, including 2 small sacs and the 3 membranous . . . . _____.

VI. PRACTICAL APPLICATIONS

Study each discussion. Then print the appropriate word or phrase in the space provided.

Group A

While observing in the emergency ward the student nurse noted the following cases.

1. Ten-year-old K had been riding his bicycle while he threw glass bottles to the sidewalk. A fragment of glass flew into one eye. Examination at the hospital showed that there was a cut in the transparent window of the eye, the . . . . . _____.

2. On further examination of K, the colored part of the eye was seen to protrude from the wound. This part is the . . . _____.

3. K's treatment included antiseptics, anesthetics and suturing of the wound. Medication was instilled in the saclike structure at the front of the eyeball. This sac is lined with a thin epithelial membrane, the .... _____.

4. A construction worker, Mr. J, was admitted because of an accident in which a piece of steel penetrated his eyeball and caused such an extensive wound that material from the inside of the eyeball oozed out. Mr. J tried to relieve the pain by forcing the jellylike material out through the wound at the front and side of his eyeball. This matter, which maintains the shape of the eyeball, is called the ... _____.

5. Because so much damage had been done, Mr. J was taken to surgery for removal of the eyeball. This operation is called .... _____.

Group B
An ear, nose and throat specialist treated the following patients one morning.

1. Mrs. B complained of some deafness and a sense of fullness in her outer ear. Examination revealed that the wax in her ear canal had hardened and formed a plug of (scientific name) .... _____.

2. Mr. J, aged 72, complained of gradually increasing deafness although he had no symptoms of pain or other problems related to the ears. Examination revealed that his deafness was the type called nerve deafness. The cranial nerve that carries impulses related to hearing to the brain is called the auditory nerve or the .... _____.

3. Mrs. C complained of deafness that resembled the type from which her aunt and her mother suffered. She asked whether she could undergo surgery since she had heard that this surgical treatment was often successful. This disorder, in which bony changes prevent the stapes from vibrating normally, is called .... _____.

4. Baby L was brought in by his mother because he awakened crying, and was holding the right side of his head. He had been suffering from a cold but now he seemed to be in pain. Examination revealed a bulging red eardrum. The eardrum is also called the .... _____.

5. The cause of baby L's painful bulging eardrum was an infection of the middle ear, called .... _____.

6. In order to relieve the pressure due to the presence of pus inside the middle ear, the eardrum was cut, a procedure called .... _____.

VII. REFERENCES

Memmler, R. L., and Wood, D. L.: The Human Body in Health and Disease, ed. 4, pp. 141-155. Philadelphia, Lippincott, 1977.

Chaffee, E. E., and Greisheimer, E. M.: Basic Physiology and Anatomy, ed. 3, pp. 230-254. Philadelphia, Lippincott, 1974.

Lenihan, J.: Human Engineering: The Body Re-examined, pp. 59-111. New York, Braziller, 1975.

Stonehouse, B.: The Way Your Body Works, pp. 14-19. London, Beasley, 1974.

CHAPTER TWELVE

The Heart and Heart Disease

I. OVERVIEW

The ceaseless beat of the heart day and night throughout one's entire lifetime is such an obvious key to the presence of life that it is no surprise that this organ has been the subject of wonderment and poetry. When the heart stops pumping, life ceases. The cells must have oxygen and it is the heart's pumping action which propels oxygenated blood to them.

In size the heart has been compared to a *closed fist*. In location it is thought of as being on the left side although about one third is to the right of the midline. The muscular apex of the triangular heart is definitely on the left. It rests *on the diaphragm*, the dome-shaped muscle that separates the chest (thoracic) cavity from the abdominal space.

The heart of birds and mammals including man has 2 sides in which the *aerated* (so-called pure) blood and the *unaerated* (lower in oxygen) blood are kept *entirely separated* from each other. So the heart is really a *double pump* in which the 2 parts pump in unison, a vital duet. Each *side* of the heart is divided into 2 parts or *chambers* though here there is direct communication. The upper chamber in each case opens directly into the lower chamber, the ventricle. The 2 ventricles pump blood to organs of the body so their walls are much thicker than the walls of the upper chambers. The *coronary arteries* supply blood to the heart muscle itself.

II. TOPICS FOR REVIEW

1. the heart as a pump
2. structure of the heart wall
   a. endocardium
   b. myocardium
   c. pericardium
3. parts of the heart
   a. septum
   b. chambers and valves
4. the conduction system of the heart
   a. sinoatrial node; pacemaker
   b. atrioventricular node
   c. atrioventricular bundle

5. normal and abnormal heart sounds
6. classification of heart disease
    a. according to tissue involved
    b. based on causative and age factors
7. congenital heart disease
8. rheumatic fever and heart injury
    a. rheumatic endocarditis
    b. vegetations and adhesions of valve parts
    c. mitral stenosis and valve incompetence
9. coronary heart disease
10. prevention of heart disease
11. instruments and medicines
12. recent devices for treating heart disorders

III. MATCHING EXERCISES

Matching only within each group, print the answer in the space provided.

Group A

arteries   mitral valve   veins
tricuspid valve   interatrial septum   aortic valve
endocardium   myocardium   interventricular septum
    pericardium   pulmonary valve

1. The membrane of which the heart valves are formed and which lines the interior of the heart is called .... *endocardium*

2. By far the thickest layer in the heart wall is the muscular one, the .... *myocardium*.

3. The outermost layer of the heart and the lining of the pericardial sac is .... *pericardium*.

4. A partition, the septum, separates the 2 sides of the heart. The thin-walled upper part of this septum is the .... *interatrial septum*

5. The larger part of the partition between the 2 sides of the heart is the .... _____.

6. Between the 2 right chambers of the heart lies the right atrioventricular valve. It is also called the .... _____.

7. The left atrioventricular valve is thicker and heavier than the right; it is made of 2 flaps or cusps. It is called the ... *mitral valve*

8. Situated between the right ventricle and the pulmonary artery is the valve that prevents blood on its way to the lungs from returning to the right ventricle. This is the ... *pulmonary valve*.

9. The valve that prevents blood from returning after the left ventricle has emptied itself is the .... *tricuspid valve*

142

10. Blood is pumped to the lungs and body tissues through _veins_.

11. Oxygenated blood from the lungs and deoxygenated blood from the body tissues is carried through the .... _arteries_.

Group B

automaticity	sinoatrial node	bundle of His
atria	lubb, dupp	systole
atrioventricular node	venous	diastole

1. The active phase of cardiac contraction is called .... _systole_.

2. Heart muscle is capable of contracting independently of nervous control. This property is called .... _automaticity_

3. The brief resting period that follows the contraction phase of the heart cycle is .... _diastole_.

4. Impulses in the heart follow a definite sequence, beginning in the pacemaker. The pacemaker is located in the upper right atrial wall and is called the .... _sinoatrial node_

5. Next the excitation wave travels throughout the muscles of the upper heart chambers causing them to contract. These are the .... _atria_.

6. Following this the second mass of conduction tissue (located in the septum) is stimulated. This is the ...... _atrioventricular node_

7. Finally the ventricular musculature contracts in response to stimulation by the branching part of the conduction system which is the .... _bundle of His_

8. Numerous disorders may cause deviations, or changes, in normal heart sounds. These sounds may be described by the syllables .... _lubb, dubb_.

Group C

degenerative heart disease	infarct
congenital heart disease	organic murmur
endocarditis	functional murmur
myocarditis	coronary heart disease
thrombus	ischemic heart disease

1. Abnormal heart sounds are called murmurs. The type of murmur that is not associated with abnormalities of the heart is called a(n) .... _functional murmur_

2. An abnormal heart sound that is evidence of damage to the heart or its vessels is a(n) .... _organic murmur_

3. A general term that is used to describe abnormalities of the heart that have been present since birth is .... _congenital_.

4. Inflammation of heart muscle is referred to as .... _myocarditis_.

5. The type of heart disease in which the valves are damaged is called .... _____.

6. Deterioration of heart tissues, most common after the age of 45, is characteristic of a condition called .... _degenerative_.

7. A serious disorder in which the walls of the blood vessels that supply the heart muscle are involved is .... _coronary_.

8. A deficiency in blood supply to the heart muscle may destroy the muscle cells, a disorder known as .... _Ischemic_.

9. An area of dead tissue that is formed as a result of a lack of blood supply is called an .... ~~_endocarditis_~~ _infarct_.

10. The technical term for a blood clot formed within a blood vessel is .... _thrombus_.

Group D

tricuspid valve	pacemaker	hypertension
fluoroscope	digitalis	anticoagulant
septum	mitral stenosis	pump-oxygenator
coronary thrombosis	stethoscope	echocardiography

1. In order to prevent the formation of a thrombus in a blood vessel the physician may prescribe a(n) .... _____.

2. A valuable drug that aids in regulating the heart beat is derived from the foxglove plant. This drug is .... _____.

3. High frequency sound vibrations are sent into the heart through the chest wall and recorded upon return by ..... _____.

4. An instrument that uses x-rays in examining deep structures is the .... _____.

5. An instrument that supplies impulses to regulate the heart beat may be implanted under the skin. It is a .... _____.

6. The simple instrument used by the physician for listening to sounds from within the patient's body is the .... _____.

7. The right atrioventricular valve is generally called the ... _____.

144

8. The 2 sides of the heart are completely separated from each other by a partition known as a(n) .... _____.

9. Many types of heart surgery can now be performed with the use of the heart-lung machine, also called the .... _____.

10. A frequent cause of enlargement of the heart is high blood pressure, or .... _____.

11. Formation of a thrombus within a heart artery may result in complete obstruction of blood flow, a condition called _____.

12. If the mitral valve becomes narrowed due to rheumatic fever, the resulting condition is .... _____.

Group E

incompetence   pericarditis   adhesions
angina pectoris   occlusion   vegetations
infection   smoking   echocardiography

1. During the course of rheumatic fever, deposits may form along the edges of the valves. These are called .... _____.

2. As the cusps thicken, they tend to stick together because of the formation of .... _____.

3. With formation of new tissue on the valve cusps, they may retract and become unable to meet as the valve attempts to close. This is known as valve .... _____.

4. Rheumatic fever develops as a result of .... _____.

5. Numerous studies indicate that the incidence of coronary heart disease is increased as much as ten times by .... _____.

6. A rapid, painless and harmless test for the presence of lesions of the heart uses sound impulses that are reflected and recorded. This is .... _____.

7. Inflammation of the serous membrane on the heart surface and the surface lining the pericardial sac is called ... _____.

8. A severe pain that is felt in the region of the heart, the left arm and the shoulder may be .... _____.

9. Complete closure of an artery is called .... _____.

IV. LABELING

For each of the following illustrations print the name or names of each labeled part on the numbered lines.

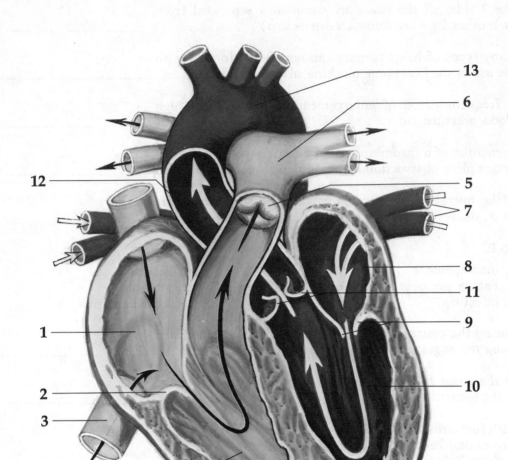

The heart and great vessels.

1. R. atrium
2. tricuspid valve
3. inferior vena cava
4. R. ventricle
5. pulmonic valve
6. pulmonary artery
7. L. pulmonary veins

8. L. atrium
9. mitral valve
10. L. ventricle
11. aortic valve
12. ascending aorta
13. aortic arch

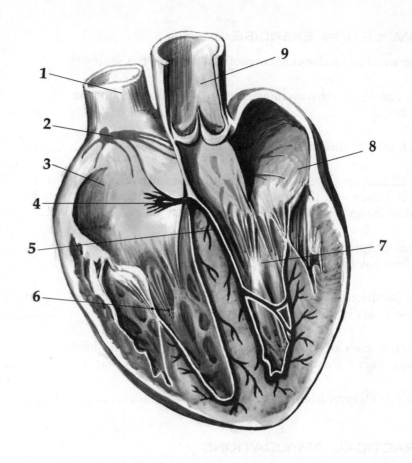

Conduction system of the heart.

1. Superior Vena Cava
2. sinoatrial node
3. R. atrium
4. atrioventricular node
5. Bundle of His
6. R. Ventricle
7. L. ventricle
8. L. atrium
9. ascending aorta

V. COMPLETION EXERCISE

Print the word or phrase that correctly completes the sentence.

1. The continuous 1-way movement of the blood is known as the .... _peristalisis_

2. Each minute the heart contracts on an average of about .. _80 pr/an._

3. Recall that one layer of serous membrane forms the lining of the closed sac; the other layer covers the organ surface. Lining the sac that encloses the heart is .... _____.

4. From Chapter 5 you learned that the serosa on the heart surface has the 2-word name .... _____.

5. The partition between the 2 thick-walled lower chambers of the heart is the .... _____.

6. Because each of the 3 parts of the 2 exit valves is half-moon shaped these valves are described as .... _____.

7. The left atrioventricular valve is called the .... _____.

VI. PRACTICAL APPLICATIONS

Study each discussion. Then print the appropriate word or phrase in the space provided.

1. Mrs. K had rheumatic fever several times during her teen-age years. Now at the age of 34 she was often short of breath and complained of some spitting-up of blood. It was found that the left atrioventricular valve had become so scarred that blood could not flow adequately from the left atrium to the left ventricle. This disorder is called ... _____.

2. Using the stethoscope to listen to Mrs. K's heart sounds, the physician detected an .... _____.

3. Mr. L was 42 years of age, and was overweight. During a game of handball he felt severe heart pains; he collapsed in shock. Examination indicated that a clot had formed in a blood vessel supplying the heart, with complete obstruction of blood flow. The scientific name for this disorder is .... _____.

4. Mr. C, age 74, had not felt well for several months. He said that he felt weak, and seemed to be out of breath after even slight exertion. Considering his history and age, his heart condition would probably be classified as ...... _____.

5. One of the first tests that were done on all these patients was a recording of electric currents produced by heart muscle. The apparatus that records this information is the _____.

6. Additional studies on Mrs. K included the introduction of a small tube into the veins of her right arm and then into the right side of her heart. This procedure is called ...... _____.

VII. REFERENCES

Memmler, R. L., and Wood, D. L.: The Human Body in Health and Disease, ed. 4, pp. 157-167. Philadelphia, Lippincott, 1977.

Chaffee, E. E., and Greisheimer, E. M.: Basic Physiology and Anatomy, ed. 3, pp. 274-288. Philadelphia, Lippincott, 1974.

Lenihan, J.: Human Engineering: The Body Re-examined, pp. 126-132. New York, Braziller, 1975.

Stonehouse, B.: The Way Your Body Works, pp. 36-37. London, Beasley, 1974.

Villee, C. A., and Delthier, V. G.: Biological Principles and Processes, pp. 561-565. Philadelphia, Saunders, 1971.

CHAPTER THIRTEEN

Blood Vessels and Blood Circulation

I. OVERVIEW

The blood vessels are classified, according to function, as *arteries*, *veins* or *capillaries*; the arteries and veins are subdivided into *pulmonary* vessels and systemic vessels.

The two arterial systems—the systemic and the pulmonary—can be likened to trees; each has a trunk, the aorta in one and the pulmonary artery in the other. Each trunk has subdivisions, large and small branches that carry the blood into the capillaries where exchanges between the blood and the tissue fluid occur. The tissue fluid provides for the transfer of substances required by the cell in exchange for those not needed or those manufactured for use elsewhere. The venous systems consist of tributaries progressing in size from small to large; they return the blood to the heart, which pumps it into the arterial trunks, thus completing the circuit.

The *pulse* rate and the *blood pressure* are manifestations of the circulation; they tell the trained person a great deal about the overall condition of the individual being examined.

II. TOPICS FOR REVIEW

1. arteries: structure and function
2. veins: structure and function
3. capillaries: structure and function
4. pulmonary vessels
5. systemic vessels
6. branches of aorta
   a. ascending
   b. aortic arch
   c. thoracic
   d. abdominal
   e. iliac
   f. other parts of arterial tree
7. anastomoses

8. systemic veins
   a. superficial
   b. deep
   c. superior vena cava
   d. sinuses
   e. inferior vena cava
9. portal circulation
10. capillaries
11. pulse
12. blood pressure
13. disorders of blood vessels
14. replacement of arteries

III. MATCHING EXERCISES

Matching only within each group, print the answer in the space provided. One answer may be used more than once and there may be some words that are not used at all.

Group A

systemic	endothelium	pulmonary
arteries	celiac	carotid
aorta	capillaries	coronary

1. The vessels that are related to the lungs, including the arteries and their branches in the lungs and the veins that drain lung capillaries are all designated as .... *Pulmonary*.

2. Exchanges between the blood and the cells take place through the .... *capillaries*.

3. Since their function is to carry blood from the heart's pumping chambers, the thickest walls are those of the blood vessels called .... *arteries*.

4. The innermost tunic of the artery is composed of .... *endothelium*.

5. The largest artery in the body is divided into 4 regions. This vessel is the .... *Aorta*.

6. The arteries that carry food and oxygen to the tissues of the body are classified as .... *Systemic*.

7. The ascending aorta has 2 branches that supply the heart muscle. Because they form a crown around the base of the heart they are classified as .... *coronary*.

8. Supplying the head and neck on each side is an artery named the .... *carotid*.

152

9. One of the unpaired arteries that supplies some of the viscera of the upper abdomen is a short trunk, the ...... *celiac*.

Group B

phrenic artery   brachial artery   common iliac arteries
brachiocephalic   lumbar arteries   left common carotid artery
    trunk   right subclavian   superior mesenteric
anastomosis   artery   artery
renal arteries   hepatic artery

1. Coming off the aortic arch is a short artery formerly called the innominate artery. This is the .... *Brachiocephalic trunk*

2. Supplying the left side of the head and neck is the ..... *(l) Common carotid*

3. Oxygenated blood is carried to the liver by the .... *hepatic artery*.

4. The largest branch of the abdominal aorta supplies most of the small intestine and the first half of the large intestine. This branch is the .... *Superior Mesenteric artery*

5. The muscular partition between the abdominal and thoracic cavities is supplied by a right and a left .... *phrenic artery*.

6. The artery supplying the arm is a continuation of the axillary artery, and is called the .... *Brachial artery*.

7. Blood supply to the right upper extremity is through the *Lumbar arteries*

8. The largest of the paired branches of the abdominal aorta are those that supply the kidneys. These are the .... *Renal arteries*

9. The abdominal aorta finally divides into 2 .... *iliac arteries*

10. A communication between 2 arteries is called an .... *anastomosis*.

11. Supply to the abdominal wall is through the .... *iliac arteries*

Group C

brachiocephalic trunk   radial artery   unpaired
mesenteric   femoral artery   basilar artery
paired   circle of Willis   volar arch
celiac trunk

1. An anastomosis of the 2 internal carotid arteries and the basilar artery is located immediately under the center of the brain. It is called the .... *circle of Willis*

2. The inferior mesenteric is an example of an artery that is *paired*.

153

3. The radial and ulnar arteries in the hand anastomose to form the .... _Valar Arch_

4. Anastomoses between branches of the vessels supplying blood to the intestinal tract comprise arches named ..... _Mesenteric arches_

5. The right subclavian artery and the right common carotid artery are branches of the .... _Brachiocephalic trunk_

6. The left gastric artery and the splenic artery are 2 of the 3 branches of the .... _Cliac trunk_

7. The union of the 2 vertebral arteries forms the .... _Basilar artery_

8. The external iliac arteries extend into the thigh. Here each of them becomes a .... _femoral_.

9. The popliteal arteries are examples of the many blood vessels that are .... _unpaired_.

10. The branch of the brachial artery that extends down the forearm and wrist of the thumb side is the .... _Radial_.

Group D

- azygos vein
- inferior vena cava
- jugular veins
- liver
- median cubital
- portal vein
- brachiocephalic veins
- saphenous vein
- superior vena cava
- venous sinuses

1. The longest vein is the superficial one called the .... _Saphenous vein_

2. Because of its location near the surface at the front of the elbow one of the veins frequently used for removing blood for testing is the .... _____.

3. The areas supplied by the carotid arteries are drained by the .... _jugular veins_.

4. The union of the jugular and subclavian veins forms the ... _____.

5. Veins draining the head, the neck, the upper extremities and the chest all empty into the .... _Superior Vena Cava_

6. Before reaching the superior vena cava (and then the heart), blood from the chest wall drains into the .... _azygos vein_.

7. Structures other than veins also drain deoxygenated blood. These structures are called .... _Venous sinuses_.

8. The blood from the parts of the body below the dia-phragm is drained by the large vein called the .... *Inferior Vena Cava*

9. Tributaries from the unpaired organs empty into a vein that enters the liver where it subdivides into smaller veins. This unusual vein is called the .... *Portal Vein*

10. Food products are released into the circulatory system from the .... _____.

Group E

sinusoids   hepatic veins   capillary walls
coronary sinus   lateral sinuses   cavernous sinuses
common iliac veins   superior sagittal sinus   gastric veins
portal tube   left testicular vein   veins
superior mesenteric vein

1. The inferior vena cava begins with the union of the 2 .... _____.

2. The only exceptions to the rule that paired veins empty directly into the vena cava are the left ovarian vein and the _____.

3. Unpaired veins coming mostly from the digestive tract are drained by a special vein called the .... _____.

4. Among the paired veins that empty directly into the inferior vena cava are those draining the liver, the ...... _____.

5. The vein that drains most of the small intestine and the first part of the large intestine is the .... _____.

6. The tributaries of the portal tube include those that drain the stomach, the .... _____.

7. Within the liver, there are no capillaries; instead, this function is performed by .... _____.

8. Cell nutrients pass into the tissue fluid through the ..... _____.

9. The veins of the heart wall drain mainly into the .... _____.

10. The ophthalmic veins drain into the .... _____.

11. Nearly all the blood from the veins of the brain eventually empties into one or the other of the transverse or ...... _____.

12. Metabolic waste products proceed through capillary walls into the blood of the capillary and then into the .... _____.

13. In the midline above the brain and in the fissure between the 2 cerebral hemispheres is a long blood-containing space called the .... _____.

Group F

atherosclerosis	dorsalis pedis	radial artery
slower	hypertension	sphygmomanometer
hypotension	faster	pulse
arteriosclerosis	aneurysm	

1. Beginning at the heart and traveling along the arteries is a wave of increased pressure started by the force of ventricular contractions. This wave is called the........... _____.

2. The wave is readily felt at the wrist because of the artery that passes over the bone on the thumb side. This is the _____.

3. Sometimes it is necessary to use the artery on the top of the foot for obtaining the pulse. This is the .... _____.

4. Blood pressure is recorded by the..................... _____.

5. It is important to recognize factors that may influence pulse rate. Emotional disturbance, for example, may cause the pulse rate to be................................. _____.

6. An abnormal decrease in blood pressure, as may occur in shock, is called .... _____.

7. Kidney disease is one cause of abnormally high blood pressure, or .... _____.

8. Weakness of a blood vessel wall may give rise to a(n).... _____.

9. As the child matures, his pulse rate normally becomes... _____.

10. A change in the arterial walls in which yellow, fatlike material replaces muscle and elastic connective tissue leads to a diagnosis of............................... _____.

11. The condition in which calcium salts and fibrous connective tissues infiltrate the artery walls and cause hardening of the arteries is called............................... _____.

Group G

shock	femoral artery	hypertension
cerebral artery	diastolic pressure	systolic pressure
facial artery	brachial artery	saphenous vein
albumin	coronary artery	

1. Headaches, dizziness, and mental disorders may be the result of sclerosis of a................................ _____.

2. A decrease in the size of artery lumens throughout the body may be a cause of................................ _____.

3. Arteriosclerosis may involve the arteries that supply the kidneys, in which case one important symptom is the appearance in the urine of the protein.................. _____.

4. Sclerosis may involve the heart vessels, in which case it affects a .... _____.

5. A diseased portion of an artery can be replaced with a blood vessel from the patient's own body. Often a large vessel from the thigh is used, namely the................ _____.

6. During ventricular relaxation the sphygmomanometer measures .... _____.

7. The most serious immediate problem following an accident usually is hemorrhage. By pressing certain arteries against the underlying bone, it is often possible to stop this hemorrhage. Hemorrhage around the nose and mouth may be stopped by pressing against the lower jaw to compress the .... _____.

8. Hemorrhage from the forearm, wrist and hand may be stopped by pressing along the groove between the 2 large arm muscles to compress the........................ _____.

9. Hemorrhage of the lower extremity may be stopped by pressing in the groin to compress the.................. _____.

10. A sudden failure of the circulation is called............. _____.

11. Such signs as cold, clammy skin, very low blood pressure, extreme pallor and shallow breathing suggest the presence of .... _____.

12. During heart muscle contraction one may measure the... _____.

IV. LABELING

For each of the following illustrations print the name or names of each labeled part on the numbered lines.

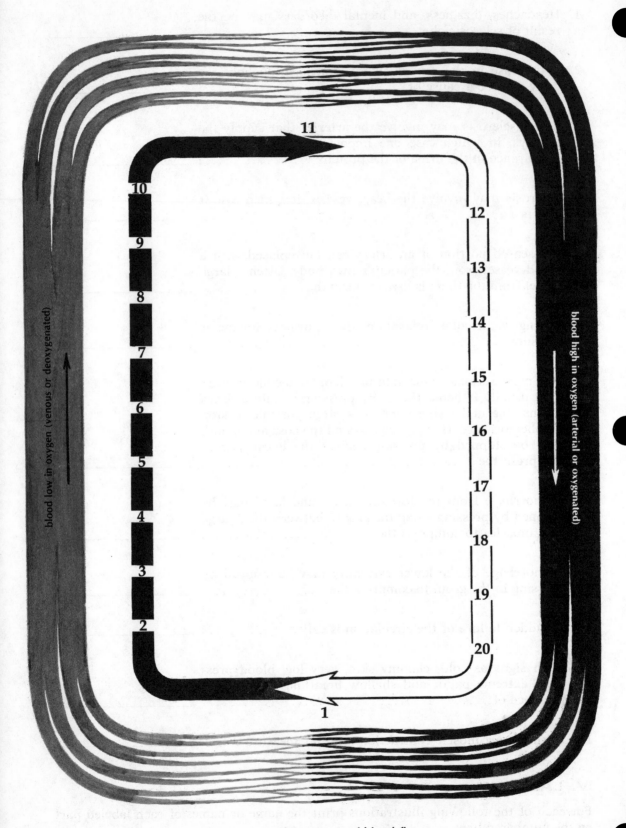

blood low in oxygen (venous or deoxygenated)

blood high in oxygen (arterial or oxygenated)

Diagram to show circuit of blood flow.

1. _____
2. _____
3. _____
4. _____
5. _____
6. _____
7. _____
8. _____
9. _____
10. _____
11. _____
12. _____
13. _____
14. _____
15. _____
16. _____
17. _____
18. _____
19. _____
20. _____

1. _____   4. _____   7. _____

2. _____   5. _____   8. _____

3. _____   6. _____

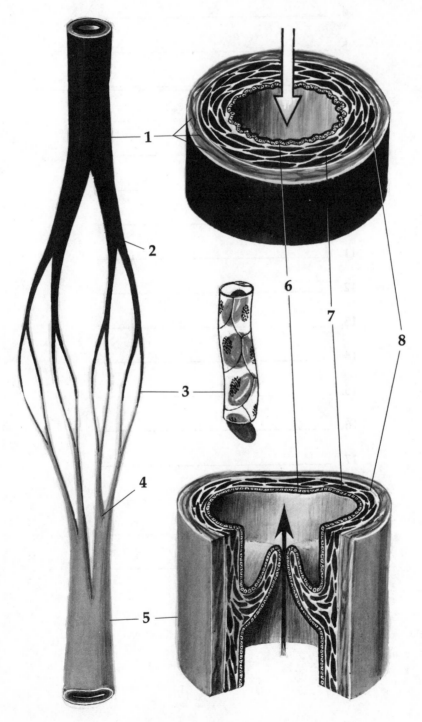

Sections of small blood vessels.

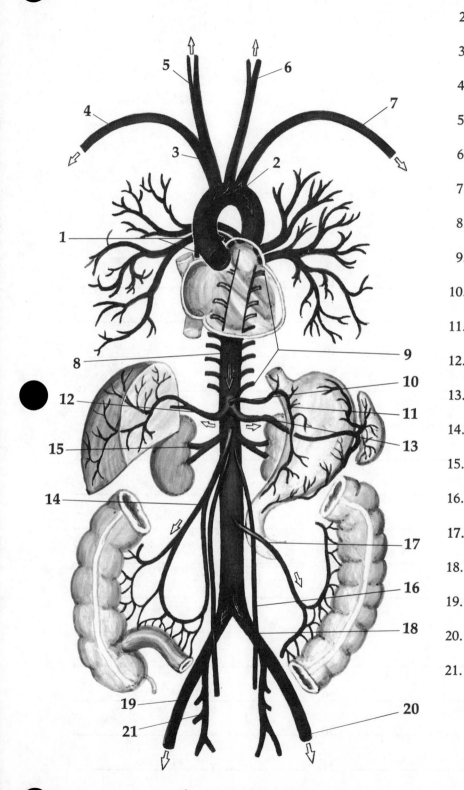

1. _____
2. _____
3. _____
4. _____
5. _____
6. _____
7. _____
8. _____
9. _____
10. _____
11. _____
12. _____
13. _____
14. _____
15. _____
16. _____
17. _____
18. _____
19. _____
20. _____
21. _____

The aorta and its branches.

1. _____   17. _____

2. _____   18. _____

3. _____   19. _____

4. _____   20. _____

5. _____   21. _____

6. _____   22. _____

7. _____   23. _____

8. _____   24. _____

9. _____   25. _____

10. _____   26. _____

11. _____   27. _____

12. _____   28. _____

13. _____   29. _____

14. _____   30. _____

15. _____   31. _____

16. _____

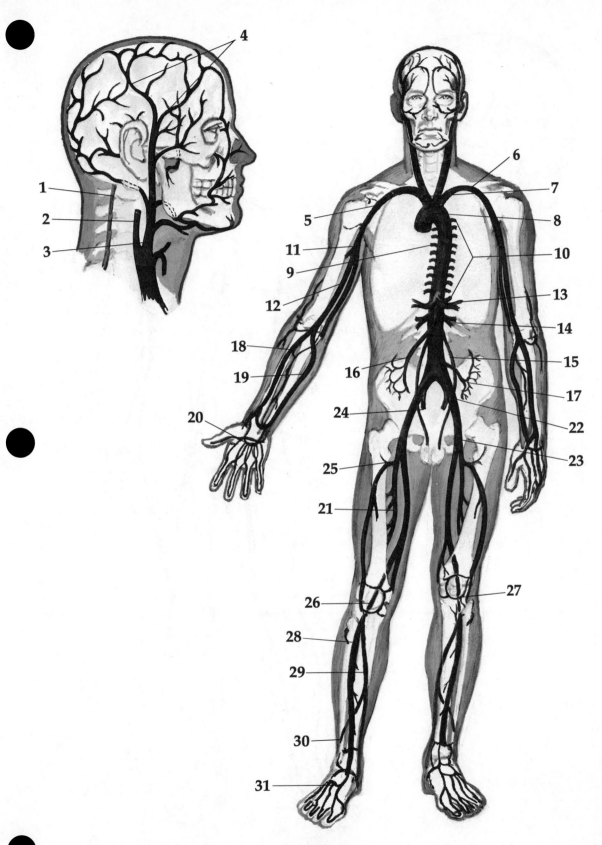

Principal arteries.

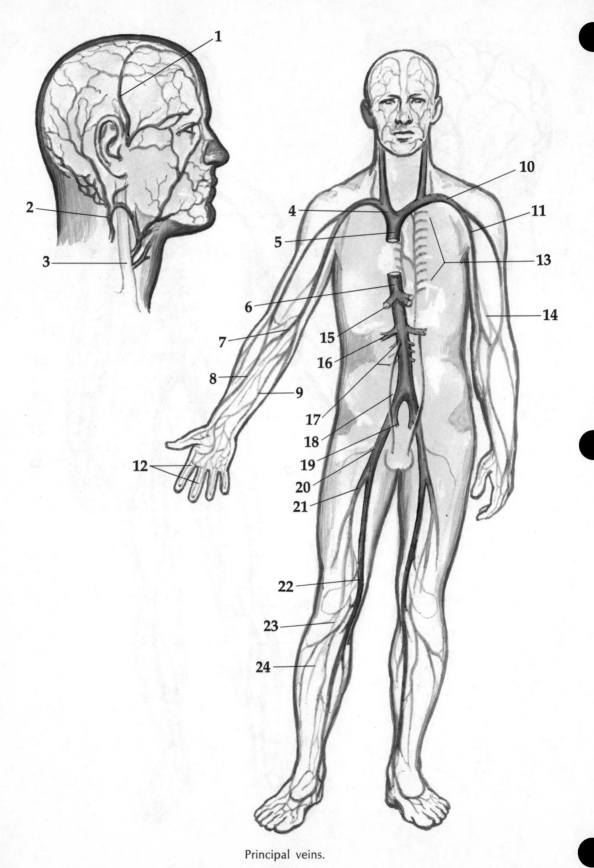

Principal veins.

1. _____

2. _____

3. _____

4. _____

5. _____

6. _____

7. _____

8. _____

9. _____

10. _____

11. _____

12. _____

13. _____

14. _____

15. _____

16. _____

17. _____

18. _____

19. _____

20. _____

21. _____

22. _____

23. _____

24. _____

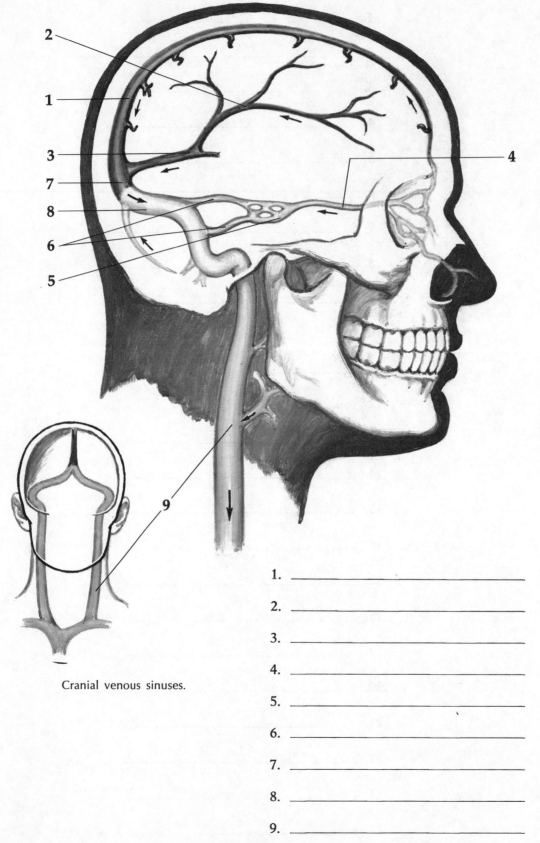

Cranial venous sinuses.

1. _____

2. _____

3. _____

4. _____

5. _____

6. _____

7. _____

8. _____

9. _____

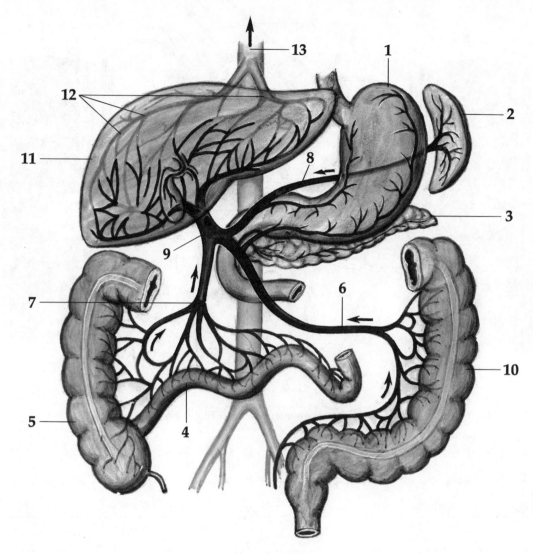

Portal circulation.

1. _____   8. _____

2. _____   9. _____

3. _____   10. _____

4. _____   11. _____

5. _____   12. _____

6. _____   13. _____

7. _____

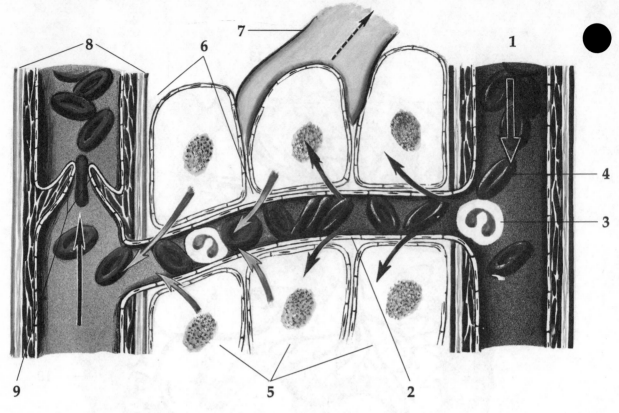

Diagram showing the connection between the small blood vessels through capillaries.

1. _____   6. _____

2. _____   7. _____

3. _____   8. _____

4. _____   9. _____

5. _____

V. COMPLETION EXERCISE

Print the word or phrase that correctly completes the sentence.

1. Deoxygenated blood is carried from the right ventricle by the .... _____.

2. The smallest subdivisions of arteries have thin walls in which there is little connective tissue and relatively more muscle. These vessels are................................ _____.

3. To supply nutrients to body tissues and carry off waste products from the tissues are functions of the circulation described as .... _____.

4. The middle tunic of the arterial wall is composed of elastic connective tissue plus......................... _____.

5. Endothelium comprises the tunic of the arterial wall that is .... _____.

6. Swollen and ineffective veins are described as.......... _____.

7. Hemorrhoids are swollen and tortuous veins located in the _____

8. Persons whose work requires them to stand much of the time frequently suffer from varicosities of the.......... _____.

9. The smallest veins are formed by the union of capillaries. These tiny vessels are called........................ _____.

10. The circle of Willis is formed by a union of the internal carotid arteries and the basilar artery. Such a union of end arteries is called an............................. _____.

VI. PRACTICAL APPLICATIONS

Study each discussion. Then print the appropriate word or phrase in the space provided.

1. Mr. S, age 53, complained of shortness of breath, weakness and pain in the left chest. Examination indicated that the left semilunar valve was not functioning properly. This valve guards the entrance into the largest artery which is the .... _____.

2. Mrs. K, age 69, was admitted to the hospital because she had fainted several times and was unable to recall events before and after these episodes. The physician diagnosed her condition as hardening of the arteries, or.......... _____.

3. In such cases as Mrs. K's the gradual narrowing of the arteries in the brain is associated with a reduction in the volume of blood passing through them. This is called _____.

4. When the blood supply to an organ is inadequate, the cells of that organ gradually die. Cell death is called.......... _____.

5. Miss J, age 78, complained of pain and swelling in the area of her saphenous vein. The term for venous inflammation is .... _____.

6. Further study of Miss J's illness indicated that a blood clot had formed in one vein. This condition is called..... _____.

7. Miss J was transferred to the intensive care unit because it was feared that the venous clot might become dislodged and be carried in the blood to her lungs. If this should happen, death would be due to................. _____.

8. Advances in medicine and surgery have made possible the replacement of damaged parts of arteries by using a segment of the saphenous vein from the patient's own body or by the use of a synthetic vessel as a................. _____.

9. Mr. B, age 67, had been diabetic for the past several years. He had neglected his diet and was careless about following his doctor's orders. Now the doctor found it necessary to order amputation of his right foot, because necrosis of the involved tissue eventually resulted in............... _____.

VII. REFERENCES

Memmler, R. L., and Wood, D. L.: The Human Body in Health and Disease, ed. 4, pp. 169-187. Philadelphia, Lippincott, 1977.

Chaffee, E. E., and Greisheimer, E. M.: Basic Physiology and Anatomy, ed. 3, pp. 292-320. Philadelphia, Lippincott, 1974.

Ehrlich, P. R., Holm, R. W., and Soulé, M. E.: Introductory Biology, pp. 306-313. New York, McGraw-Hill, 1973.

Lenihan, J.: Human Engineering: The Body Re-examined, pp. 122-125. New York, Braziller, 1975.

Villee, C. A., and Delthier, V. G.: Biological Principles and Processes, pp. 565-569. Philadelphia, Saunders, 1971.

CHAPTER FOURTEEN

The Lymphatic System and Lymphoid Tissue

I. OVERVIEW

Lymph is the watery fluid that flows within the lymphatic system. It originates from the blood plasma and from the tissue fluid that is found in the minute spaces around and between the body cells. Lymph may contain certain cellular waste products as well as fat globules (from the digestive system following a meal). The relatively few cells that are present are usually lymphocytes. The fluid moves from the *lymphatic capillaries* through the *lymphatic vessels* and thence to the *right lymphatic duct* and the *thoracic duct*. The lymphatic vessels are thin-walled and delicate; like some veins, they have valves that prevent backflow of tissue fluid.

The *lymph nodes*, which are the system's filters, are composed of *lymphoid tissue*. These nodes remove small foreign bodies such as dead blood cells, carbon particles and pathogenic organisms; they also manufacture *lymphocytes* and *antibodies*. Chief among them are the *cervical nodes* in the neck, the *axillary nodes* in the armpit, the *tracheobronchial nodes* near the trachea and bronchial tubes, the *mesenteric nodes* between the peritoneal layers, and the *inguinal nodes* in the groin area.

In addition to these are several organs of lymphoid tissue whose functions are somewhat different. The *tonsils* filter tissue fluid. The *thymus* is essential for antibody formation and development of immunity during the early weeks of life. The *spleen* has numerous functions, among which are the destruction of used-up blood cells, as a reservoir for blood, and the production of red cells before birth.

A distinctive attribute of lymphoid tissue is its abundance. Many more of these masses are normally present than are actually needed, so that removal of certain of them does not interfere with the overall functioning of the human organism.

II. TOPICS FOR REVIEW

1. lymphatic system
   a. lymph

(1) composition
(2) function
b. lymph conduction
2. lymphoid tissue
a. location
b. functions
c. 5 main groups of lymph nodes
d. other lymphoid structures differing in function from main groups
3. disorders of lymphatic system and lymphoid tissue

III. MATCHING EXERCISES

Matching only within each group, print the answer in the space provided.

Group A

right lymphatic duct	blood	inguinal nodes
lacteals	valves	chyle
endothelium	buboes	axillary nodes
cervical nodes		

1. There is easy passage of soluble materials and water through the walls of lymphatic capillaries, in which a single layer of cells forms the......................... _____.

2. The lymphatics resemble some veins in that they contain structures that prevent backflow. These are............. _____.

3. One pathway for fats from digested food to the blood-stream is through specialized lymphatic capillaries of the intestine that are called .... _____.

4. Lymph is drained from the right side of the head, of the neck, of the thorax and of the right upper extremity by the _____.

5. The combination of fat globules and lymph gives rise to a milky-appearing fluid called .... _____.

6. Lymph nodes are named according to location. Those located in the armpits are known as.................. _____.

7. Drainage of lymph from the lower extremities and the external genitalia is through the........................ _____.

8. Abnormally large inguinal nodes, as may be found in certain infections, are called.................................. _____.

9. The final destination of filtered lymph is the............. _____.

10. The lymph nodes located in the neck and draining certain parts of the head and neck are known as................ _____.

Group B

thymus   pharyngeal tonsils   bile

lingual tonsils   lymphangitis   spleen

spleen   adenitis   palatine tonsils

1. The oval lymphoid bodies located at each side of the soft palate are known as................................ _____.

2. The enlarged masses of lymphoid tissue often found on the back wall of the pharynx and commonly called adenoids are correctly called............................ _____.

3. At the back of the tongue are masses of lymphoid tissue called .... _____.

4. The structure that is believed to be essential in the formation of antibodies very early in life is the.............. _____.

5. Blood filtration is carried out by an organ located in the upper left quadrant (left hypochondriac region) of the abdomen. This is the................................ _____.

6. The end products of red blood cell destruction are returned to the liver, where they are manufactured into a secretion called .... _____.

7. During embryonic and fetal life red blood cells are produced in the .... _____.

8. A word that actually means inflammation of a gland is used to refer to inflammatory disorders of lymph nodes, structures that are not glands. This word is............. _____.

9. Recall that the suffix -itis means inflammation; when referring to lymph vessel inflammation, we use the word _____.

Group C

lymph nodes   backflow   radial lymphatic vessels

phagocytosis   drainage   lymph

subclavian vein   antibodies   macrophages

thoracic duct

1. Located in the bone marrow, the spleen and lymph nodes are the cells that absorb and destroy foreign matter. These are .... _____.

2. An important function of lymphoid tissue is the production of chemical substances that aid in combatting infection. These are called................................ _____.

3. The spleen generates cells that are able to engulf bacteria and other foreign cells. This process is known as......... _____.

4. The fluid that moves from blood plasma to the tissue spaces and finally to special collection vessels is called... _____.

5. Before the lymph reaches the veins it is passed through organs that act as filters. These are.................... _____.

6. The lymphatic vessels serve as a system for............. _____.

7. Lymph received in the right lymphatic duct drains into the right .... _____.

8. Lymph is drained from the body below the diaphragm and on the left side above the diaphragm by the largest lymphatic vessel, the .... _____.

9. The valves of the lymphatic vessels prevent lymph....... _____.

10. Lymphatic vessels are named according to location; thus, those on the lateral side of the forearm are the.......... _____.

Group D

lymphocytes	hilum	plasma
lymph	lacteals	lymph nodes
veins	blindly	cisterna chyli

1. Lymphatic capillaries differ from blood capillaries in that they begin .... _____.

2. The first part of the thoracic duct is enlarged, forming a temporary storage area. It is called the................ _____.

3. Chyle, the fluid formed by combination of lymph and fat globules, comes from the intestinal.................... _____.

4. An important function of lymph nodes is the manufacture of white blood cells known as......................... _____.

5. Intercellular fluid originates from the liquid part of the blood. The liquid part of the blood is called............. _____.

6. Tissue fluid passes from the intercellular spaces into the lymphatic vessels; it is then called..................... _____.

7. The masses of lymphoid tissue that filter foreign substances from the liquid lymph are known as............. _____.

8. Both superficial and deep vessels are found in the lymphatic system just as in the case of the system of........ _____.

174

9. The area of exit for the vessels carrying lymph out of the node is known as the................................ _____.

IV. LABELING

Print the name or names of each labeled part on the numbered lines.

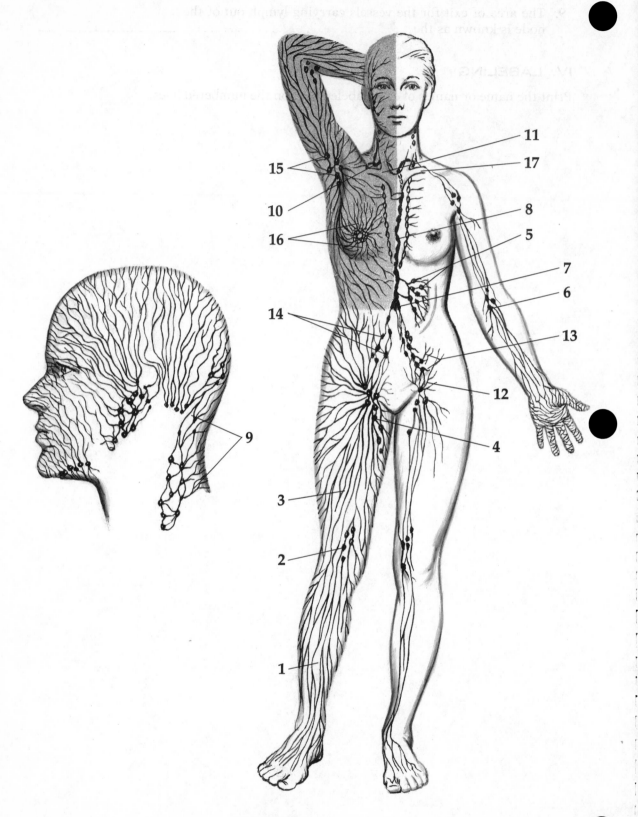

The lymphatic system.

1. _____   9. _____

2. _____   10. _____

3. _____   11. _____

4. _____   12. _____

5. _____   13. _____

6. _____   14. _____

7. _____   15. _____

8. _____   16. _____

                        17. _____

V. COMPLETION EXERCISE

Print the word or phrase that correctly completes the sentence.

1. Since the prefix angi- means blood or lymph vessel, it is to be expected that the term for inflammation of lymph vessels is .... _____.

2. Lymphatic vessels from the left side of the head, neck, and thorax empty into the largest of the lymph vessels, the _____.

3. The lymph from the body below the diaphragm and from the left side above the diaphragm is carried into the blood of the .... _____.

4. Between the 2 layers of peritoneum that form the mesentery are found nodes called............................ _____.

5. In city dwellers nodes may appear black because they become filled with carbon particles. This is true mostly of the nodes that surround the windpipe and its divisions. These are the .... _____.

6. A disease prevalent during the Middle Ages was responsible for the death of thousands of people. This disease was characterized by the presence of buboes—inflammatory swellings of the inguinal nodes—so it was called.... _____.

7. The structure popularly known as adenoids is correctly called the .... _____.

177

8. The spleen and other organs produce cells that can engulf harmful bacteria and other foreign cells, by a process called .... _____.

9. The blockage of lymphatic vessels by filariae may cause tremendous enlargement of the lower extremities, a disorder called .... _____.

10. A tumor that occurs in lymphoid tissue, whether benign or malignant, has the general name of................. _____.

VI. PRACTICAL APPLICATIONS

1. Mrs. B, age 38, underwent biopsy of a small mass in her right breast which was positive for cancer. She was now being admitted in order to have a radical mastectomy. In this operation certain nodes are removed as well as the breast, because cancer cells from the breast often invade them. These are the armpit nodes called the............. _____.

2. Mr. G, age 31, complained of swellings in his neck, his armpits, his groin and other areas. A diagnosis of Hodgkin's disease was made. Treatment with radiation and antineoplastic drugs was planned. The nodes of the neck are designated the .... _____.

3. Mr. K, age 41, had been hunting wild rabbits in the central valley of California. Several days after dressing a number of these rabbits an ulcer developed on his hand. A tentative diagnosis of tularemia, or rabbit fever, was made. The infecting organisms had been carried to the axillary nodes via tubes called....................... _____.

4. Mrs. M was admitted for study because her spleen was enlarged. This condition is called...................... _____.

5. After a number of tests had been done, a diagnosis of splenic anemia was made. Removal of the spleen was carried out. This operation is called a.................. _____.

6. Mr. L was admitted for study of pelvic masses. It was feared that his disorder might be a malignance of lymphoid tissue that is rapidly fatal, called................ _____.

7. Mr. J was 21 years old. His complaint concerned swelling in the groin region. A blood test showed that the young man had contracted syphilis. Infection of the external genitalia is often followed by the appearance of buboes, which are enlarged .... _____.

VII. REFERENCES

Memmler, R. L., and Wood, D. L.: The Human Body in Health and Disease, ed. 4, pp. 189-196. Philadelphia, Lippincott, 1977.

Chaffee, E. E., and Greisheimer, E. M.: Basic Physiology and Anatomy, ed. 3, pp. 321-327. Philadelphia, Lippincott, 1974.

Nason, A., and Dehaan, R. L.: The Biological World, pp. 419-420. New York, Wiley, 1973.

Schmidt-Nielson, K.: Animal Physiology, pp. 146-148. New York, Cambridge University Press, 1975.

Villee, C. A., and Delthier, V. G.: Biological Principles and Processes, pp. 569-571. Philadelphia, Saunders, 1971.

CHAPTER FIFTEEN

Digestion and Indigestion

I. OVERVIEW

The complex process by which the food we eat reaches the cells throughout the body is accomplished through *digestion* and *absorption*. These are functions of the *digestive system*; its components are the *alimentary canal* and the *accessory organs*.

The alimentary canal, consisting of the *mouth*, the *pharynx*, the *esophagus*, the *stomach* and the large and small *intestine*, forms a continuous passageway in which ingested food is prepared for utilization by the body, and waste products are collected to be expelled from the body. The *liver*, the *gallbladder*, and the *pancreas*, which comprise the accessory organs, manufacture various substances needed to regulate food metabolism, serve as storage areas for certain substances which are released as needed, and function in other ways to help maintain a normal state of health.

Since ingested food is the main source of nourishment for the body, a *balanced diet*, with avoidance of "fads in foods," should be followed by all persons who are in a normal state of health.

II. TOPICS FOR REVIEW

1. components of the digestive system
2. oral cavity
3. peristalsis
4. swallowing tubes and accessories
5. stomach
6. small intestine
7. large intestine
8. liver
9. gallbladder
10. pancreas
11. peritoneum
12. disorders of alimentary canal structures
13. disorders of accessory organs
14. nutrition and diet

III. MATCHING EXERCISES

Matching only within each group, print the answer in the space provided. Some answers may be used more than once.

Group A

tongue	permanent teeth	deciduous teeth
incisors	alimentary canal	molars
absorption	digestion	premolars

1. The process by which ingested food is converted into substances that may be taken into the cells is known as...... _____.

2. The transfer of digested food to the bloodstream is called _____.

3. The structures and organs through which ingested food or its breakdown products pass comprise the............. _____.

4. The mouth, the pharynx, the esophagus, the stomach, and the intestine are part of the.......................... _____.

5. One can differentiate taste sensations by means of special organs of the .... _____.

6. The baby molars are replaced by permanent teeth called bicuspids or .... _____.

7. The grinding teeth located in the back part of the oral cavity are called .... _____.

8. The temporary baby teeth are lost, and are therefore described as .... _____.

9. During the time that the baby teeth are appearing, or erupting, the second set of teeth are developing in the jawbones. These are the............................... _____.

10. The first 8 of the baby teeth to appear are the........... _____.

11. The wisdom teeth usually appear during the later teen years. They are more accurately described as the third... _____.

12. The cutting teeth located in the front portion of the buccal cavity are the................................. _____.

Group B

first molars	third molars	32 teeth
second molars	premolars	20 teeth
oral cavity	canines	

1. The baby molar teeth are not replaced by permanent molars but by smaller......................... _____.

2. Decay and infection of baby molar teeth may easily spread to the first permanent teeth, the............... _____.

3. The so-called eyeteeth are the.......................... _____.

4. The incisors are located in the front part of the......... _____.

5. Normally, at about age 12 the jawbones should be large enough to accommodate the erupting................. _____.

6. It sometimes happens that the jaw is not large enough to accommodate the last teeth to erupt. These teeth are identified as the .... _____.

7. By the time the baby is 2 years old he should have all the deciduous teeth. This means there are................. _____.

8. The molar teeth are found in the back portion of the..... _____.

9. An adult who has a full set of permanent teeth has...... _____.

Group C

parotitis	pharynx	alveolus
saliva	pyorrhea alveolaris	mastication
esophagus	stomatitis	spirochete
gingivitis	deglutition	uvula
caries	peristalsis	

1. Infection of the gum is known as...................... _____.

2. Infection of the mucous membrane lining of the mouth, except for the gums, is called.......................... _____.

3. Certain inflammations involve the tooth pocket, or....... _____.

4. An inflammation of the tooth socket which is associated with a discharge of pus is called....................... _____.

5. Loss of teeth is often the result of tooth decay, or dental _____.

6. The process of chewing is called....................... _____.

7. The act of swallowing is known as...................... _____.

183

8. An essential part of the digestive process involves the coating of the food with mucus and the dissolving of the food in the mouth by the digestive juice............ _____.

9. The common disorder, mumps, is referred to medically as _____.

10. Food and liquid are prevented from entering the nasal cavities by the soft fleshy............................ _____.

11. Food is propelled along the alimentary canal by the rhythmic motion known as............................ _____.

12. The tongue pushes the food into the.................... _____.

13. The contents of the alimentary canal are moved along as far as the stomach by the gullet, or.................... _____.

14. In an earlier chapter pathogenic microorganisms were described. One of these is the causative organism of Vincent's angina. It is a.............................. _____.

Group D

cardiac valve
pyloric sphincter
gastric juice
rugae

pharynx
chyme
flatulence

soft palate
epiglottis
vomiting

1. The uvula hangs from the back of the roof of the oral cavity. This part of the oral cavity roof is the .... _____.

2. During deglutition there is contraction of the muscles of the .... _____.

3. With the muscular contraction that occurs during deglutition, the openings into the air spaces above and below the mouth are closed off by the soft palate and by the.... _____.

4. The structure that guards the entrance into the stomach is called the .... _____.

5. The valve between the distal end of the stomach and the small intestine is the.................................. _____.

6. If the stomach is empty, there will be many folds in the lining. These folds are called .... _____.

7. Contents in the stomach are mixed with hydrochloric acid and enzymes to form .... _____.

8. The combination of hydrochloric acid and enzymes in the stomach is referred to as .... _____ .

9. The feeling of illness known as nausea may be followed by reverse peristalsis, resulting in .... _____ .

10. Excessive air (gas) in the stomach or intestine may cause considerable discomfort and a condition referred to as ... _____ .

Group E

adenocarcinoma	villi	ileocecal
hypoacidity	peptic ulcer	sugars
pyloric stenosis	ileum	fats
jejunum	hyperacidity	duodenum

1. An abnormally high production of stomach acid is called _____ .

2. Persistent indigestion is an important warning of possible cancer of the stomach. The most frequent type, which originates from the stomach lining, is called .... _____ .

3. The first part of the small intestine is the .... _____ .

4. An abnormally small opening out of the stomach may be due to a constriction of the sphincter. This condition is called .... _____ .

5. Disintegration of tissue associated with loss of membrane substance may occur in many areas of the body; when occurring on the mucous membrane of the esophagus, the stomach or the duodenum it is called .... _____ .

6. The absorbing area of the small intestinal mucosa is greatly increased by numerous projections called .... _____ .

7. Lying just beyond the duodenum is the second part of the small intestine, the .... _____ .

8. Although bile contains no enzymes, it aids in the digestion of .... _____ .

9. The final, and longest, section of the small intestine is the _____ .

10. Because of its location the valve between the small and large intestine is described as .... _____ .

11. An abnormally low production of stomach acid may be an indication that serious illness is present. Such underacidity is called .... _____.

12. Among the classes of nutrients that are essential to cell life are carbohydrates, which include .... _____.

Group F

glycogen	hepar	hepatitis
heparin	trypsin	lipase
fibrinogen	mesentery	amylopsin
ascites		

1. Nerves, arteries and other structures supplying the small intestine are found between the 2 layers of peritoneum called the .... _____.

2. Should fluid accumulate in the peritoneal cavity, as may occur in certain serious illnesses, the condition is called .. _____.

3. The largest gland in the body is the liver, or .... _____.

4. Inflammation of the liver is called .... _____.

5. The liver has many essential functions. One of these is the manufacture of a substance which prevents clotting of the blood. This is .... _____.

6. Sugar is stored by the liver and released as simple sugar (glucose) as needed. The form in which sugar is stored is _____.

7. The liver has many functions including the production of plasma proteins such as albumin and .... _____.

8. Pancreatic juice contains enzymes that act in various ways on the chyme in the small intestine. Starch is changed to sugar by the pancreatic enzyme .... _____.

9. Fats must be broken down into simpler compounds in order to be readily absorbed. The enzyme responsible for this breaking down is .... _____.

10. Proteins enter the bloodstream in the form of amino acids. The splitting of proteins is accomplished by the pancreatic enzyme .... _____.

Group G

liver	lacteals	fecal matter
ptyalin	ducts	small intestine
cecum	vermiform appendix	colon

1. The important function of absorption is carried out through the numerous villi projecting from the mucosa of the .... _____ .

2. Following their absorption into the bloodstream through the capillary walls of the villi, food materials are stored and released as needed by the .... _____ .

3. In the saliva there is an enzyme that begins starch digestion. It is called .... _____ .

4. Much of fat absorption occurs through the lymphatic capillaries of the villi; they are called .... _____ .

5. Digestive juices are carried from accessory organs of digestion to the duodenum by means of .... _____ .

6. The materials to be eliminated will continue through the ileocecal valve into the beginning of the large intestine. Here it enters a small pouch called the .... _____ .

7. The small blind tube attached to the proximal part of the large intestine is called the .... _____ .

8. In the large intestine, layers of involuntary muscle move the solid waste products on toward the rectum. This waste material is called .... _____ .

9. The longer part of the large intestine is the .... _____ .

IV. LABELING

For each of the following illustrations print the name or names of each labeled part on the numbered lines.

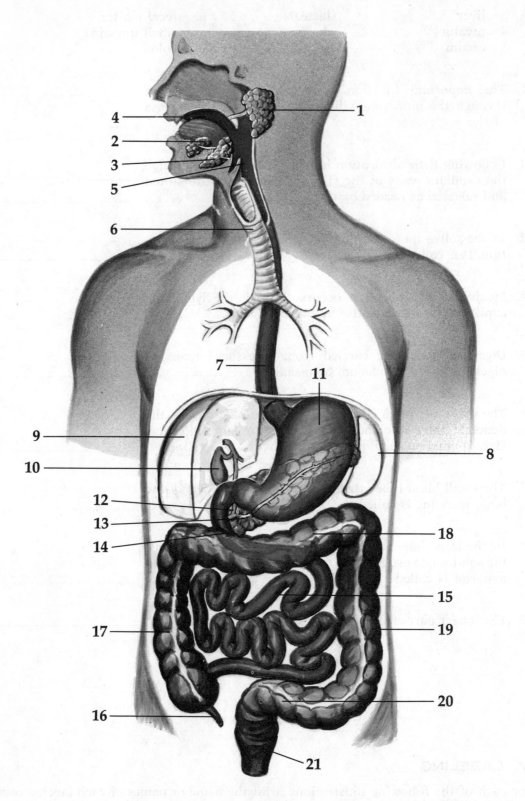

The digestive system.

1. _____
2. _____
3. _____
4. _____
5. _____
6. _____
7. _____
8. _____
9. _____
10. _____
11. _____

12. _____
13. _____
14. _____
15. _____
16. _____
17. _____
18. _____
19. _____
20. _____
21. _____

1. _____
2. _____
3. _____
4. _____
5. _____

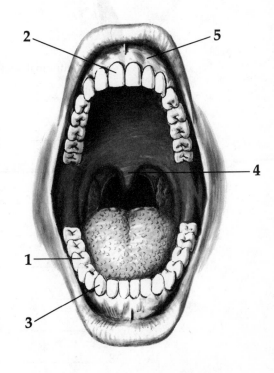

The mouth, showing teeth and tonsils.

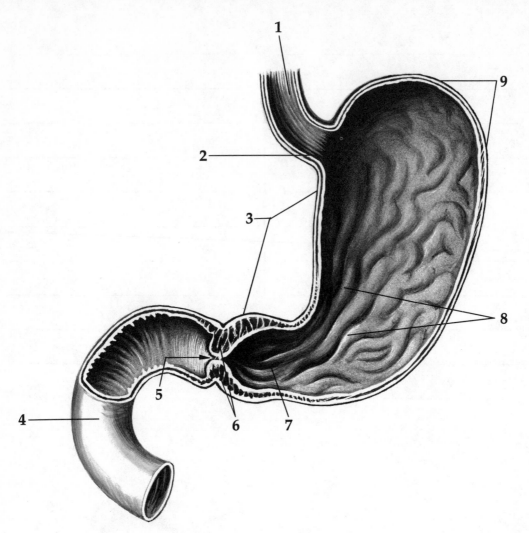

Longitudinal section of the stomach.

1. _____
2. _____
3. _____
4. _____
5. _____
6. _____
7. _____
8. _____
9. _____

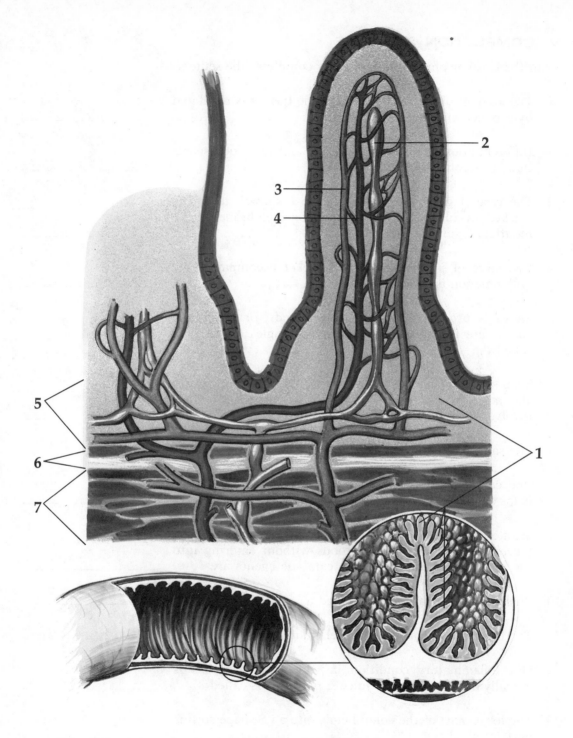

Structure of a villus.

1. _____   5. _____

2. _____   6. _____

3. _____   7. _____

4. _____

191

V. COMPLETION EXERCISE

Print the word or phrase that correctly completes the sentence.

1. The amount of sugar "burned" in the tissues is regulated by a pancreatic hormone called .... _____.

2. Differentiation of bitter, salty, sweet and sour food sensations is accomplished through the .... _____.

3. The type of gingivitis that is due to a spirochete and is marked by ulceration of the mucous membrane of the mouth and gums is called .... _____.

4. Discharge of pus from a tooth socket accompanied by inflammation is called .... _____.

5. Saliva is produced by 3 pairs of glands, of which the largest are the ones located near the angles of the jaw, called the .... _____.

6. Such symptoms as nausea, vomiting, diarrhea and severe abdominal pain are characteristic of an inflammation that involves the stomach and bowels called .... _____.

7. The salivary glands located under the tongue are the .... _____.

8. One component of gastric juice kills bacteria and thus helps defend the body against disease. This substance is .. _____.

9. Most of the digestive juices contain substances that cause the chemical breakdown of foods without entering into the reaction themselves. These catalytic agents are ..... _____.

10. Starches and sugars are classified as .... _____.

11. The highest concentration of fuel value is provided by ... _____.

12. One inflammatory condition in which the pancreas is actually destroyed by the juice it produces is called ..... _____.

13. The lower part of the colon bends into an S-shape so this part is called the .... _____.

14. A temporary storage section for indigestible and unabsorbable waste products of digestion is a tube called the _____.

15. The portion of the distal part of the large intestine called the anal canal leads to the outside through an opening called the .... _____.

16. When the intestinal musculature is overstimulated, as by nervous tension, the intestinal lumen may become too small to permit passage of fecal material. This condition is called .... _____.

17. The muscular sac in which bile is stored to be released as needed is called the .... _____.

18. In the United States many problems related to nutrition result from use of the wrong foods. This condition is called .... _____.

19. A stone lodging in the gallbladder duct would be likely to cause .... _____.

VI. PRACTICAL APPLICATIONS

Study each discussion. Then print the appropriate word or phrase in the space provided.

1. Mr. C, age 36, complained of pain in the "pit of the stomach." Ingestion of food seemed to provide some relief. The physician ordered x-ray studies to be done. These studies indicated that the first part of the small intestine was involved. This short section is called the ... _____.

2. Mr. C was a tense man who felt it was important to do well in his business; he worked long hours. Because of this constant stress, his physician suspected that excessive hydrochloric acid was being produced. The x-ray films confirmed the presence of tissue destruction which is characteristic of .... _____.

3. Three-month-old John was brought to the clinic by his mother because he had suffered several bouts of vomiting and could not retain food. The tentative diagnosis was a constricted or spastic pyloric sphincter, a condition called _____.

4. Mrs. K, age 24, complained that since returning from a brief trip out of the United States she was suffering from frequent watery stools. This symptom is called .... _____.

5. It was determined, through various studies, that Mrs. K had an infestation by the organism *Entamoeba histolytica*. This results in a disease called .... _____.

6. Mrs. D, age 47, complained of "indigestion" associated with pain under the ribs of her right side. Tests revealed the presence of stones in the gallbladder. The scientific name for this disorder is .... _____.

193

7. Further study of Mrs. D's case revealed episodes of gall-bladder infection, or .... _____.

VII. REFERENCES

Memmler, R. L., and Wood, D. L.: The Human Body in Health and Disease, ed. 4, pp. 197-213. Philadelphia, Lippincott, 1977.

Chaffee, E. E., and Greisheimer, E. M.: Basic Physiology and Anatomy, ed. 3, pp. 378-413. Philadelphia, Lippincott, 1974.

Lenihan, J.: Human Engineering: The Body Re-examined, pp. 152-162. New York, Braziller, 1975.

Stonehouse, B.: The Way Your Body Works, pp. 28-35. London, Beasley, 1974.

CHAPTER SIXTEEN

Respiration

I. OVERVIEW

Oxygen is supplied to the tissue cells and *carbon dioxide* is removed from them through the arrangement of spaces and passageways known as the *respiratory system*. This system comprises the *nasal cavities*, the *pharynx*, the *larynx*, the *trachea* and the *lungs*.

The 2 phases of breathing are *inhalation*—the drawing in of air—and *exhalation*—the expulsion of air. The rate normally varies from 12 to 25 times per minute. Breathing is under the control of the *respiratory center* in the brain located in the *medulla*.

Fads abound regarding ways to secure proper ventilation of one's environment. The most suitable "rules" also are the simplest: moderate coolness to remove some body heat without chilling, avoidance of drafts and moderate air circulation to dispel pollutants.

II. TOPICS FOR REVIEW

1. inhalation and exhalation
2. external respiration
3. internal respiration
4. respiratory tract
5. thoracic cavity
6. respiratory rates; variations from normal
7. disorders of respiration
8. disorders of the respiratory tract
9. diseases of lung
10. purposes of ventilation
11. equipment for respiratory tract treatment

III. MATCHING EXERCISES

Matching only within each group, print the answer in the space provided.

Group A

cellular (or internal) respiration	larynx	trachea
nasal septum	pharynx	oxygen
external respiration	conchae	carbon dioxide

1. The word "respiration" means "to breathe again"; one of its basic purposes is to supply the body cells with the gas _____.

2. At the same time that the required gas is being supplied, another gas, a waste product of cell metabolism, is being removed. This waste gas is............................. _____.

3. The aspect of respiration involving gas exchanges in the lungs is called .... _____.

4. The second aspect of respiration refers to gas exchanges within the body cells. This is called.................... _____.

5. Below the nasal cavities is a part which is common to both the digestive and respiratory systems. This is the... _____.

6. The cartilaginous structure commonly referred to as the voice box has the scientific name of.................. _____.

7. Several parts of the respiratory tract are kept open by a framework of cartilage. One of these is the windpipe or.. _____.

8. The partition separating the 2 nasal cavities is called the _____.

9. The surface over which the air moves is increased by 3 projections located at the lateral walls of each nasal cavity. These are the................................ _____.

Group B

trachea	vascular membrane	nasolacrimal duct
hilum (or hilus)	epiglottis	vocal folds
bronchi	esophagus	sinuses
nasopharynx	oropharynx	ciliated epithelium

1. The lining of the nasal cavities contains many blood vessels and is therefore described as a................. _____.

2. The small cavities in the bones of the skull are lined with mucous membrane. They are called.................... _____.

3. Tears are carried from the lacrimal glands across the eye surface, into openings at the corner of the eye, and finally into the nasal cavities by means of a tube called the..... _____.

4. The mucous membranes lining the tubes of the respiratory system are usually made of...................... _____.

5. Immediately behind the nasal cavity is the upper portion of the muscular pharynx, the........................ _____.

6. The portion of the pharynx located behind the mouth is the .... _____.

7. The lowest part of the pharynx, the laryngeal pharynx, opens into the air passageway of the larynx, located toward the front, and into the food path, toward the back, where it enters the . . . . _____.

8. The production of speech is aided by the flow of air from the lung to vibrate the . . . . _____.

9. Food is prevented from entering the remainder of the respiratory tract by closure of the glottis during swallowing. This is accomplished by a leaf-shaped structure called the . . . . _____.

10. To conduct air to and from the lungs is the purpose of the windpipe, or . . . . _____.

11. The 2 main air tubes to the lungs, formed by division of the trachea, are the . . . . _____.

12. Each bronchus plus the blood vessels and nerves that accompany it enter the lung at a notch or depression called the . . . . _____.

Group C

diaphragm	squamous epithelium	inhalation
glottis	bronchial tree	exhalation
alveoli	mediastinum	bronchiole

1. Separating the 2 vocal cords is the . . . . _____.

2. Each bronchus subdivides into many branches. This resemblance to a tree accounts for the name, ... _____.

3. The smallest division of a bronchus is called a . . . . _____.

4. At the end of each terminal bronchiole is a cluster of air sacs, called . . . . _____.

5. Easy passage for the gases that enter and leave the blood in the alveolar capillaries is provided by the very thin air sac wall. This 1-cell layer is made of . . . . _____.

6. The heart is situated in the space between the lungs called the . . . . _____.

7. The physiology of respiration involves 2 phases of breathing. In the first phase air is drawn into the lungs. This is _____.

8. In the second phase of breathing air is expelled from the alveoli. This phase is called . . . . _____.

9. Separating the thoracic cavity from the abdominal cavity is the muscular .... _____.

Group D

dyspnea	bronchoscope	pneumonectomy
respirator	tracheotomy	tracheostomy
pulmotor	phrenicotomy	suction apparatus
exudate	tuberculosis	emphysema

1. The physician is able to inspect the bronchi and the larger bronchial tubes by using an instrument containing properly arranged mirrors. This tubular instrument is the.... _____.

2. In order to remove mucus and other substances from the respiratory tract one may use various types of........... _____.

3. When the capacity to perform the normal motions of respiration has been lost, a mechanical device may be used to pump oxygen into the lungs under pressure. Such a device is called a................................ _____.

4. In order to maintain respiratory movements without effort on the patient's part in poliomyelitis (and in other cases) it may be necessary to use an "iron lung" or...... _____.

5. An incision into the trachea for the purpose of removing a foreign object or a growth is called a................ _____.

6. A metal tube is inserted into the trachea to serve as an intake for air as well as an exhaust duct for carbon dioxide in an operation called a........................ _____.

7. In 1900 the most frequent cause of death in the United States was .... _____.

8. Chronic bronchitis is often the forerunner of a respiratory disorder that ranks high as a cause of death. It is characterized by obstruction of bronchial tubes and dyspnea. It is called .... _____.

9. Products of infection, as found in the alveoli of a pneumonia patient, form a fluid called a(n)................. _____.

10. The lung may be immobilized by cutting a nerve that supplies the diaphragm. This procedure is called a....... _____.

11. The symptom of labored or difficult breathing is called.. _____.

12. The operation employed to remove a lung in a patient with bronchogenic carcinoma is called................. _____.

IV. LABELING

Print the name or names of each labeled part on the numbered lines.

The respiratory tract.

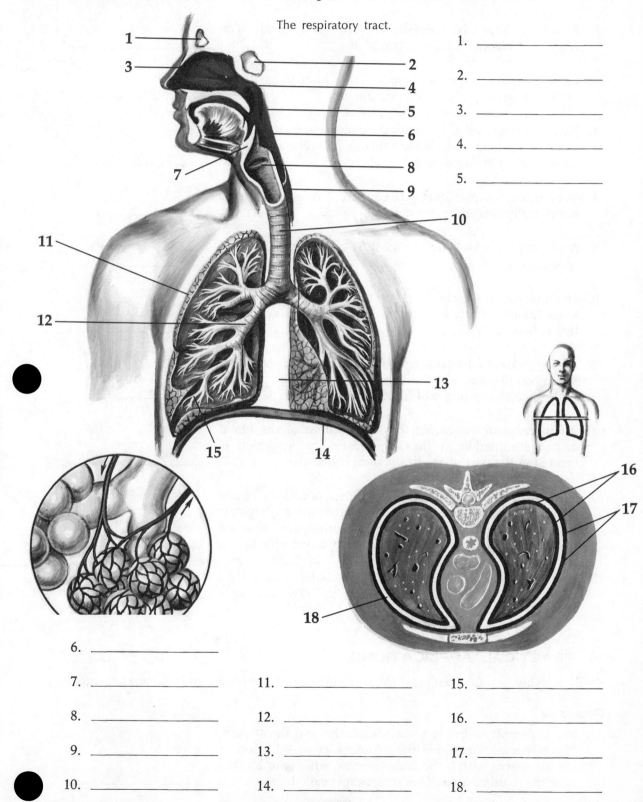

1. _____

2. _____

3. _____

4. _____

5. _____

6. _____

7. _____   11. _____   15. _____

8. _____   12. _____   16. _____

9. _____   13. _____   17. _____

10. _____   14. _____   18. _____

199

V. COMPLETION EXERCISE

Print the word or phrase that correctly completes the sentence.

1. A lack of oxygen is especially harmful to the brain. The word that means "lack of oxygen" is.................. _____.

2. Lack of tissue oxygen accompanied by increase in tissue carbon dioxide results in suffocation, or................ _____.

3. Heart disease and other disorders may cause the bluish color of the skin and visible mucous membranes characteristic of a condition called.......................... _____.

4. Very often the partition between the 2 nasal cavities is structurally defective; this defect is called............. _____.

5. An injury or a blow to the nose is a frequent cause of nosebleed, or of .... _____.

6. In emphysema the right side of the heart is overworked. Lung failure results in heart failure, a serious complication called .... _____.

7. Although great advances have been made in the control of infectious diseases, the number one killer in the communicable disease group still is......................... _____.

8. An increase in the carbon dioxide content of the blood causes stimulation of the respiratory center, which is located in the part of the brain stem called the .... _____.

9. The greatest frequency of tuberculosis occurs in conditions of overcrowding and lack of attention to hygienic measures. The tuberculosis bacillus withstands exposure to many disinfectants but it is especially vulnerable to.... _____.

10. Certain diplococci, staphylococci, chlamydias and viruses may cause an inflammation of the alveoli. The disease is _____.

VI. PRACTICAL APPLICATIONS

Study each discussion. Then print the appropriate word or phrase in the space provided.

Group A

1. Mr. C complained of a severe headache and facial pain. The physician diagnosed the problem as an infection of the air spaces within the cranial bones which are located near the nasal cavities. This infection is called........... _____.

2. D, age 5, had a profuse discharge from his nose; examination revealed inflammation of the nasal mucosa, or....... _____.

3. Mr. G, age 14, complained of a sore throat and difficulty in swallowing. The physician described the disorder as... _____.

4. Miss F, age 24, complained of hoarseness and said that it was causing her difficulty in speaking to her students. This type of inflammation is called..................... _____.

5. Mrs. D had been suffering from a cold for several days. The physician explained that no entirely effective method of prevention is known as yet. The etiologic agent is a very contagious .... _____.

6. Because of Mrs. D's lowered disease resistance, she was also suffering from inflammation of the bronchi and their subdivisions. This infection is called.................... _____.

7. The physician warned Mrs. D that if she did not stop working temporarily, and get sufficient rest in order to increase her disease resistance, her illness might extend into the lung with a resulting inflammation of the alveoli, or .... _____.

8. Mr. G, age 47, was advised to see his doctor because a routine x-ray examination performed at his place of work revealed a lung lesion. Mr. G was a chain smoker. The possibility of a malignancy of the type that originates in the lung tubes was being considered. This most common form of lung cancer is.............................. _____.

Group B

1. Mrs. S's x-ray examination revealed the presence of lung masses of the kind that form when tuberculosis bacilli invade tissue. This capsule of new tissue is called a...... _____.

2. Miss J, age 15, had had many episodes of coughing and had expectorated blood-tinged sputum. This is evidence of cavity formation in the lung. Laboratory tests may be used to identify the organism that causes tuberculosis. This bacillus is named .... _____.

3. Mr. B had neglected his health following a bout with pneumonia. There was evidence of a tuberculous infection, and fluid that accumulated in the pleural sacs. Such a collection of fluid is called an........................ _____.

4. Mr. L had been working on a cattle ranch for several years; he often drank unpasteurized milk. Examination now revealed that he had contracted tuberculosis from this milk. This type of tuberculosis is referred to as..... _____.

5. A problem that frequently develops during treatment of tuberculosis with drugs is that new strains of bacteria may cause a form of the disease that cannot be treated effectively with these particular medications. Such bacterial strains are said to be........................... _____.

6. Mrs. S, age 67, suffered from the painful condition of pleurisy, which was associated with tuberculosis in her case. The pain was caused by the rubbing together of the pleurae of the lung and the chest wall. As these surfaces stick together there is the development of.............. _____.

VII. REFERENCES

Memmler, R. L., and Wood, D. L.: The Human Body in Health and Disease, ed. 4, pp. 215-228. Philadelphia, Lippincott, 1977.
Chaffee, E. E., and Greisheimer, E. M.: Basic Physiology and Anatomy, ed. 3, pp. 350-375. Philadelphia, Lippincott, 1974.
Schmidt-Nielson, K.: Animal Physiology, pp. 37-43. New York, Cambridge University Press, 1975.
Stonehouse, B.: The Way Your Body Works, pp. 40-41. London, Beasley, 1974.

CHAPTER SEVENTEEN

The Urinary System and Its Disorders

I. OVERVIEW

The urinary system comprises 2 *kidneys*, 2 *ureters*, 1 *urinary bladder* and 1 *urethra*. This system is usually thought of as the body's main *excretory* mechanism; it is, in fact, often called the *excretory* system. The kidney, however, performs 2 other essential functions: it aids in maintaining *water balance* and in regulating *acid-base balance*.

Prolonged or serious diseases of the kidney nearly always have devastating effects on overall body function and health; a person can live without eyesight and without hearing; but life cannot be maintained unless at least 1 kidney is functioning efficiently. For this reason, renal *dialysis* and renal *transplantation* have been developed in recent years. These methods are helping to save the lives of many persons who otherwise would die of kidney failure and uremia.

II. TOPICS FOR REVIEW

1. excretory mechanisms and interrelationships
   a. digestive system
   b. respiratory system
   c. urinary system
   d. integumentary system
2. kidneys
   a. location
   b. structure
   c. functions
      (1) excretion
      (2) water balance
      (3) acid-base balance
   d. kidney replacements
3. ureters

4. urinary bladder
5. urethra
6. urine
    a. normal constituents
    b. abnormal constituents
7. disorders of the urinary system

III. MATCHING EXERCISES

Matching only within each group, print the answer in the space provided.

Group A

digestive system	kidneys	elimination
respiratory system	adipose capsule	excretion
retroperitoneal space	urine	fibrous capsule

1. Removal of waste products from the body is called...... _____.

2. By contrast, the actual emptying of the hollow organs in which waste substances have been stored is referred to as _____.

3. Other systems besides the urinary system perform excretory functions. To mention one example, bile is excreted by the .... _____.

4. The system regulating excretion of carbon dioxide and water is the .... _____.

5. The urinary system excretes water, nitrogenous waste products and salts, all of which are contained in the..... _____.

6. Extraction of wastes from the blood is a function of the _____.

7. The membranous connective tissue structure that is normally loosely adherent to the kidney itself is called the ... _____.

8. The area behind the peritoneum which contains the pancreas, duodenum and the 2 kidneys is referred to as the .. _____.

9. The circle of fat that helps to support the kidney is called the .... _____.

Group B

collecting tubules	epithelium	filtration
renal basin	Bowman's capsule	urea
cortex	convoluted tubule	reabsorption
glomerulus		

1. The cluster of capillaries located at one end of the nephron is the .... _____.

2. Materials that have passed through the capillary walls enter the first part of the nephron, the .... _____.

3. The nephron is basically a tiny coiled tube called a ..... _____.

4. The useful substances that have escaped through the nephron capillaries are sent back to the bloodstream by a process of .... _____.

5. Since the kidney is a gland, it is made up mainly of ..... _____.

6. The combination of glomerulus and Bowman's capsule connected with each of the one million nephrons of the kidney provides a highly effective means of .... _____.

7. The part of the kidney containing the nephron bulbs and their blood vessels is the .... _____.

8. Within the medulla the open ends of the nephron tubes come together (and empty into) the .... _____.

9. The urine produced by structures located in the cortex is collected by tubules in the medulla. These latter tubules empty into the pelvis, or .... _____.

10. As body cells use protein, nitrogenous waste products are produced; the chief such product is .... _____.

Group C

mineral salts	internal sphincter	peristalsis
glucose	urethra	hilum
calyces	organic	acid-base balance

1. Urine is moved along the ureter from the kidneys to the bladder by the rhythmic contraction known as .... _____.

2. Near the bladder outlet are circular muscle fibers that contract to prevent emptying. They form what is known as the .... _____.

3. The tube that carries urine from the bladder to the outside is the .... _____.

4. Because nitrogen waste products originate from living organisms they are said to be .... _____.

5. Inorganic compounds normally contained in urine are also classified as .... _____.

6. Diabetes mellitus may be suspected if a test of the urine shows the presence of the simple sugar .... _____.

7. The area where the artery, the vein and the ureter connect with the kidney is known as the .... _____.

8. Tubelike extensions that project from the renal pelvis into the kidney tissue serve to increase the area for collection of urine. These extensions are called .... _____.

9. The kidney helps prevent conditions of excessive alkalinity or acidity by regulating the body's .... _____.

Group D

edema	nephritis	nephrons
diffusion	pyelitis	salts
cystitis	buffers	calculi
acid	uremia	water balance

1. The body's intake and output of water must be carefully regulated at all times to maintain a normal state of health. The kidney aids in mantaining this .... _____.

2. The term that means inflammation of kidney tissue is .... _____.

3. A swollen, puffy appearance of the involved area may be due to an excessive accumulation of fluid in the body tissues. This condition is called .... _____.

4. All types of kidney inflammation are characterized by destruction of the microscopic units, or .... _____.

5. When the kidneys are unable to remove poisonous substances from the blood, these substances accumulate and may cause the serious condition called .... _____.

6. When uric acid or calcium salts precipitate out of the urine instead of remaining in solution, stones are formed. These are called .... _____.

7. Inflammation of the urinary bladder is called .... _____.

8. Stagnation of urine due to interference with its normal flow may cause inflammation of the kidney basin, or .... _____.

9. In order to remove nitrogen waste products from the blood, dialysis utilizes the principle of .... _____.

10. Exhalation of carbon dioxide is one means by which there is removal of substances that are .... _____.

11. Acids are neutralized by alkalis to form .... _____.

12. The nearly neutral state of the blood is due to the presence of the mineral salts known as .... _____.

IV. LABELING

For each of the following illustrations print the name or names of each labeled part on the numbered lines.

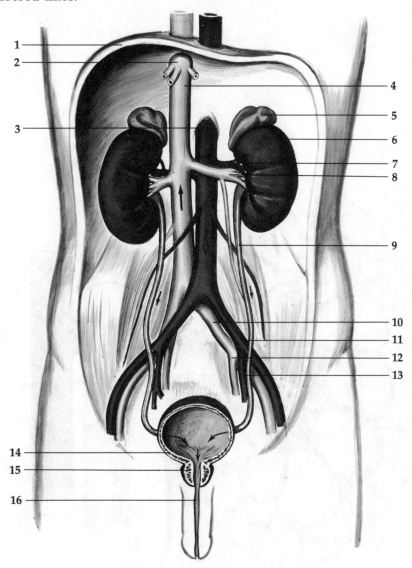

Urinary system with blood vessels.

1. _____   7. _____   13. _____

2. _____   8. _____   14. _____

3. _____   9. _____   15. _____

4. _____   10. _____   16. _____

5. _____   11. _____

6. _____   12. _____

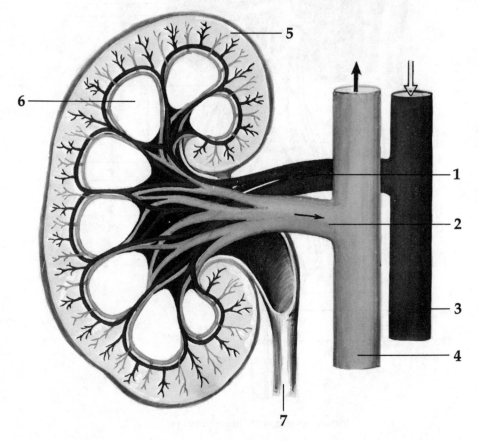

Blood supply and circulation of kidney.

1. _____   5. _____

2. _____   6. _____

3. _____   7. _____

4. _____

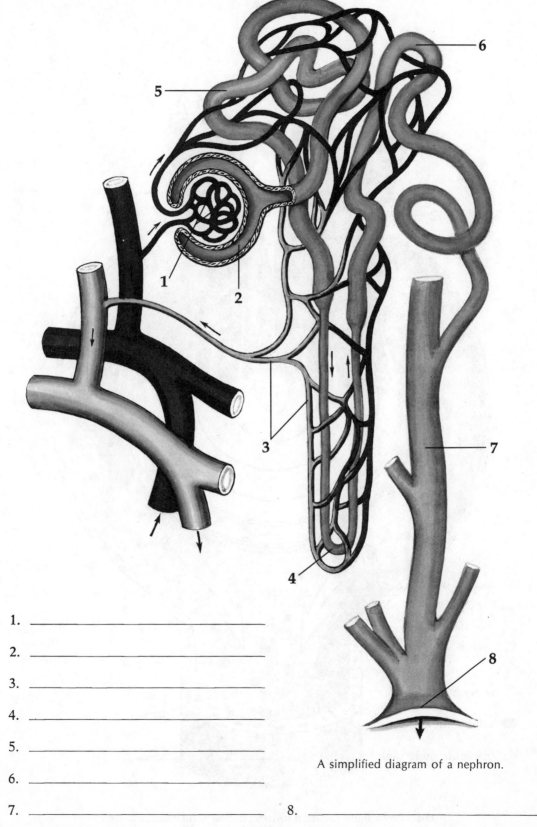

A simplified diagram of a nephron.

1. _____

2. _____

3. _____

4. _____

5. _____

6. _____

7. _____   8. _____

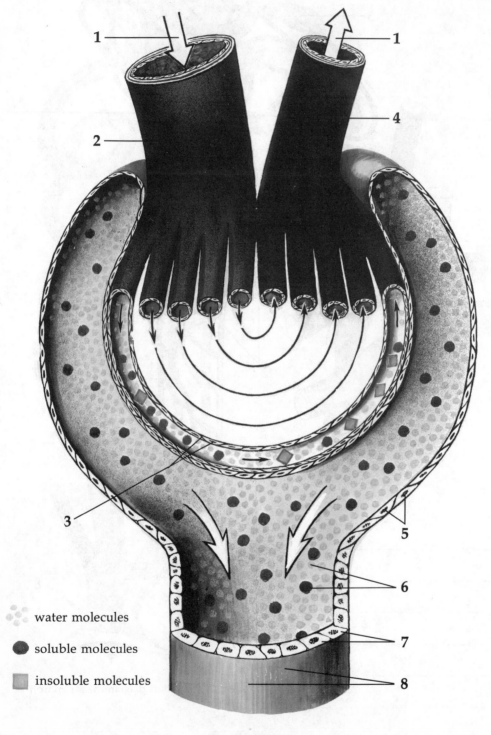

water molecules

soluble molecules

insoluble molecules

Diagram to show filtration process during formation of urine.

1. _____   5. _____

2. _____   6. _____

3. _____   7. _____

4. _____   8. _____

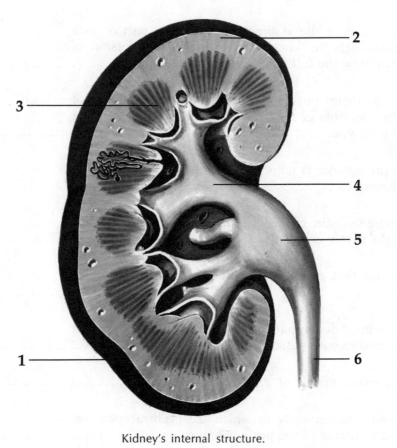

Kidney's internal structure.

1. _____   4. _____

2. _____   5. _____

3. _____   6. _____

V. COMPLETION EXERCISE

Print the word or phrase that correctly completes the sentence.

1. An inflammation of the renal pelvis which extends into the kidney tissue results in .... ———————.

2. Kidney stones that are large enough to fill the kidney pelvis and extend into the calyces are called .... ———————.

3. Inflammation of the mucous membrane and the urethral glands is characteristic of .... ———————.

4. The ureters are located behind and, at the lower part, below the peritoneum; that is, they are .... ———————.

5. Because of the oblique direction of the last part of each ureter through the lower bladder wall, compression of the ureters by the full bladder prevents .... ———————.

6. There are many causes of ureteral obstruction. One of these is a kinking of the ureter that is due to a dropping of the kidney, or .... ———————.

7. When the bladder is empty, its lining is thrown into the folds known as .... ———————.

8. The extreme pain caused by passage of a small stone along the ureter is called .... ———————.

9. The vessel that carries oxygenated blood to the kidney is the .... ———————.

10. In the male, 2 ducts that carry sex cells join the first part of the urethra as it passes through the .... ———————.

11. Urine consists mainly of .... ———————.

12. The congenital anomaly in which the urethra opens on the under surface of the penis is known as .... ———————.

13. The straddle type of injury occurs when, for example, a man is walking along a raised beam and slips so that the beam is between his legs. Such an accident may rupture the .... ———————.

14. An important sign of urinary tract disease or injury is blood in the urine, a condition that is called .... ———————.

15. In diabetes, starvation, and other conditions, fats are not completely oxidized. A test of the urine may reveal the presence of .... _____ .

16. One indication of nephritis is the presence in the urine of molds that have been formed in the kidney tubules. They are called .... _____ .

VI. PRACTICAL APPLICATIONS

Study each discussion. Then print the appropriate word or phrase in the space provided.

1. An x-ray examination of Mrs. L's kidneys revealed destruction of the calyces which gave them a moth-eaten appearance. The cause was believed to be a blood-borne infection that may have originated in the lungs, the bones, or the lymph nodes, namely .... _____ .

2. Mrs. L's history noted that her father had died of Bright's disease; the term more widely used today is .... _____ .

3. Mr. R had been exposed to arsenical substances in his work as foreman in a chemical manufacturing plant. Tests now indicated a degenerative process involving the epithelium of the convoluted tubules. The resulting disease is known as .... _____ .

4. In order to remove the accumulated urea and other nitrogenous waste products Mr. R was treated by .... _____ .

5. The x-ray examination of Miss G's urinary tract revealed a structural abnormality of the ureter in the form of extreme narrowing, or .... _____ .

6. Mrs. K was suffering from cystitis, or bladder infection. Studies indicated that there was relaxation of the pelvic floor causing stagnation of urine in the bladder, and corrective surgery was planned. In preparation for this, a catheter was inserted into the external opening, the ..... _____ .

7. Mr. K, age 61, required several studies to determine the cause of obstruction of his urinary tract. Pus cells were found in his urine specimen. This is an important sign of _____ .

8. One of the studies done in Mr. K's case revealed that there was an obstruction at the bladder neck, a disorder that is fairly common in men of his age. The obstruction was caused by enlargement of the gland through which the first part of the urethra passes. This is the .... _____ .

213

9. Mrs. C, age 38, was hospitalized because she was seriously ill and required intensive study. She had suffered from several streptococcal and staphylococcal infections. Now there was such extensive kidney damage that consideration was being given to a recently developed operation in which the kidney of a healthy person replaces the diseased kidney in the sick person. This procedure is called _____.

10. Mr. G, age 58, had not consulted a physician for many years. Now he came into the hospital seriously ill. Physical examination revealed enlarged kidneys and ureters. Laboratory studies showed that there was a severe uremia. Despite all measures, Mr. G died within a few days. Autopsy disclosed the presence of greatly distended pelves and calyces on both sides. This condition is known as ... _____.

11. Further studies at Mr. G's autopsy revealed a grossly enlarged prostate gland which had almost completely obstructed the outflow of urine. The resulting continuous back pressure over many years had caused a wasting of the kidney tissue, that is, an .... _____.

VII. REFERENCES

Memmler, R. L., and Wood, D. L.: The Human Body in Health and Disease, ed. 4, pp. 229-242. Philadelphia, Lippincott, 1977.
Chaffee, E. E., and Greisheimer, E. M.: Basic Physiology and Anatomy, ed. 3, pp. 435-468. Philadelphia, Lippincott, 1974.
Lenihan, J.: Human Engineering: The Body Re-examined, pp. 139-151. New York, Braziller, 1975.
Stonehouse, B.: The Way Your Body Works, pp. 42-43. London, Beasley, 1974.

CHAPTER EIGHTEEN

Glands and Hormones

I. OVERVIEW

Glands are organs that manufacture secretions. The glands are divided into 2 types: the *exocrine* glands have ducts that carry the secretion to other parts of the body; the *endocrine* glands are ductless, and the blood and the lymph carry their secretions. The secretions are also divided into 2 classes: the *external* secretions are carried from the gland cells to a nearby organ or to the body surface; the *internal* secretions are carried to all parts of the body by the blood or the lymph.

The endocrine glands manufacture *hormones*, chemical substances that regulate the activities of various body organs. These hormones perform many essential functions, of which control of *body growth*, control of *food metabolism* and control of the growth and development of the *sexual organs* are examples. So important, in fact, is the work of the hormones that they are often referred to as the body's *chemical messengers*.

Among the hormones that are being studied intensively at the present time are the neurohormones and the prostaglandins. Much remains to be learned about them. At present about all that is known is that they help to regulate various chemical reactions within the body, but not very much is understood about their effects.

II. TOPICS FOR REVIEW

1. secretions
   a. external
   b. internal
2. glands
   a. exocrine
   b. endocrine
3. hormones
   a. thyroid gland; thyroxine
   b. parathyroid gland; parathormone
   c. pituitary gland; anterior lobe, posterior lobe
   d. pancreas
   e. adrenal glands
   f. sex glands

4. placenta
5. thymus
6. pineal body
7. medical uses of hormones

III. MATCHING EXERCISES

Matching only within each group, print the answer in the space provided. The same answer may be used more than once.

Group A

iodine	basal metabolism	islands (or islets) of
external secretions	hormones	Langerhans
parathyroid glands	thyroid	suprarenal glands
	medulla	cretinism

1. The digestive juices and tears are examples of . . . . _____ .

2. The substances produced by endocrine glands are known as . . . . _____ .

3. Glands that have ducts produce substances classified as . . _____ .

4. The body's "chemical messengers" are the . . . . _____ .

5. The adrenal glands are also known as the . . . . _____ .

6. The groups of specialized cells scattered throughout the pancreas are known as the . . . . _____ .

7. The largest of the endocrine glands is located in the neck. It is the . . . . _____ .

8. The adrenal glands and the kidneys have an inner part called the . . . . _____ .

9. An individual born without functioning thyroid tissue is said to suffer from . . . . _____ .

10. One test for evaluating thyroid function is done when the person is at complete rest. This test determines the person's . . . . _____ .

11. Located behind the thyroid gland and embedded in its capsule are the 4 . . . . _____ .

12. A relatively simple test for thyroid function involves taking blood from a vein and then testing it for the so-called protein-bound . . . . _____ .

Group B

pituitary	myxedema	adrenal
goiter	placenta	insulin
thyroxine	iodine	parathormone

1. Production of heat and energy in the body tissues is regulated by the hormone . . . . _____.

2. In order that thyroxine may be manufactured, the blood must contain an adequate supply of. . . . . _____.

3. Enlargement of the thyroid gland results in a swelling of the neck called . . . . _____.

4. Atrophy of the thyroid in the adult causes mental and physical sluggishness; the term used to describe this condition is . . . . _____.

5. The amount of calcium dissolved in the circulating blood is regulated by the parathyroid secretion. . . . . _____.

6. Several essential hormones are produced by the anterior and posterior lobes of the. . . . . _____.

7. In order to provide for normal sugar utilization in the tissues, the islands of Langerhans must produce the hormone _____.

8. The external cortex and the internal medulla act as separate glands with specific functions in the case of the. . . . . _____.

9. The normal development of the embryo is aided by hormones from the ovaries, pituitary, and an organ present only during pregnancy, namely the. . . . . _____.

10. In the disorder known as diabetes mellitus, sugar is not "burned" in the tissues to produce energy. This is due to a lack of the hormone. . . . . _____.

Group C

aldosterone	oxytocin	exophthalmic goiter
endocrine	estrone	ACTH
cortisol	exocrine	progesterone
epinephrine		lymphocytes

1. Great nervousness, a rapid pulse, weight loss and bulging of the eyes are symptoms of Graves' disease, or. . . . . _____.

2. During stressful situations such as injury or surgery the body is protected somewhat by a hormone (a glucocorticoid) that is usually called. . . . . _____.

3. The thymus produces hormones that stimulate the production of cells needed in the body's defense against infection. These cells are the........................... _____.

4. Contraction of the pregnant uterus is stimulated by a hormone from the posterior pituitary called............. _____.

5. Blood pressure is raised and the rate of the heart beat is increased by the chief hormone of the adrenal medulla... _____.

6. The reabsorption of sodium in the kidney tubules is an important electrolyte-regulating function of the adrenal cortex hormone .... _____.

7. Testosterone is produced by the male sex glands; the female sex glands produce a hormone which most nearly parallels testosterone in its action. This hormone is called _____.

8. The lacrimal gland is an example of one that is described as .... _____.

9. A hormone that is necessary for normal development of pregnancy is one produced by the female sex glands. It is called .... _____.

10. When the needs of the body are such that amino acids must be changed to sugar instead of protein, the adrenal cortex produces large amounts of the hormone.......... _____.

11. The pituitary is stimulated by impulses from the hypothalamus, while the adrenal cortex is stimulated by the anterior pituitary hormone known as.................. _____.

12. The somatotropic hormone is an example of one of many hormones that are manufactured by glands that are classified as .... _____.

IV. LABELING

Print the name or names of each labeled part on the numbered lines.

1. _____   6. _____

2. _____   7. _____

3. _____   8. _____

4. _____   9. _____

5. _____

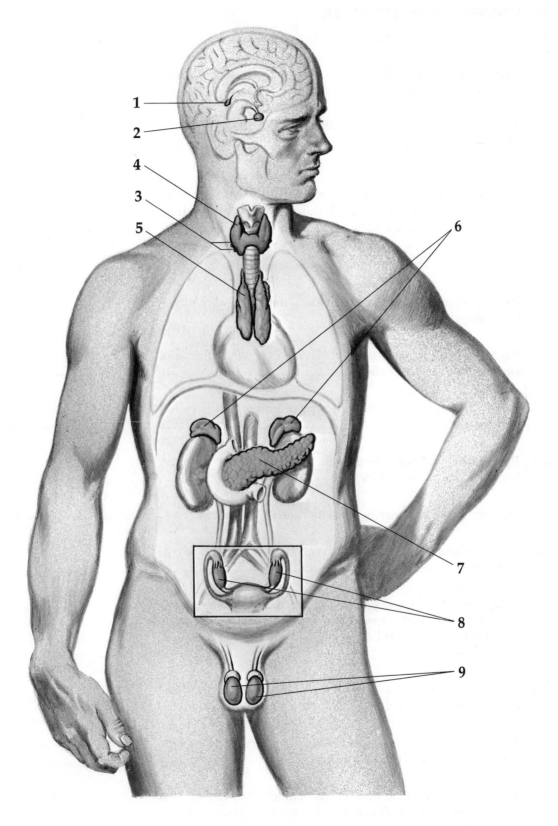

Endocrine system.

V. COMPLETION EXERCISE

Print the word or phrase that correctly completes the sentence.

1. Endocrine glands lack the means to carry secretions from the glands; hence they are said to be................. _____.

2. In addition to the connective tissue framework, most glands, whether exocrine or endocrine, are made largely of .... _____.

3. Hormones perform many essential functions that have been compared with those of the nervous system. In a sense hormones are chemical......................... _____.

4. The 2 lobes of the thyroid are connected by the.......... _____.

5. In myxedema, the person becomes mentally and physically sluggish because of thyroid...................... _____.

6. Because it is common to people living in a particular area simple goiter is also called............................ _____.

7. Tumor formation is found in a type of goiter that is called nodular goiter or .... _____.

8. Epinephrine, which affects the blood pressure and the heart rate, is produced by the......................... _____.

9. The term used to describe an underactive thyroid is...... _____.

10. Graves' disease, in which the basal metabolic rate is tremendously increased, is a common form of............. _____.

11. Should production of parathormone decrease, there will be a decrease in the amount of calcium dissolved in the blood. This may be followed by muscle spasms, a condition called .... _____.

12. A person born with a lack of the growth-promoting hormone will remain a............................... _____.

13. An abnormal increase in production of the hormone epinephrine may result from a tumor of the............... _____.

14. The hormone that stimulates the smooth muscle of the blood vessels is produced by the part of the pituitary called the .... _____.

15. The hypothalamus stimulates the anterior pituitary to produce ACTH, which in turn stimulates hormone production by the .... ——————————.

16. Cretinism is due to lack of the hormone.................. ——————————.

VI. PRACTICAL APPLICATIONS

Study each discussion. Then print the appropriate word or phrase in the space provided.

1. Mr. J, age 23, required evaluation of pituitary function. As part of this evaluation, an x-ray examination was planned because of the possibility that a tumor was the cause of his excessive height of 7 feet as well as his abnormal weakness. The tests revealed that a pituitary tumor was present, so his condition was diagnosed as............. ——————————.

2. Seventeen-year-old Miss K had never had a menstrual period. The cause could have been a deficiency of the ovarian hormone called .... ——————————.

3. Mrs. C, age 56, had been brought to the hospital in coma, that is, she was unconscious and could not be aroused. Tests revealed that her blood sugar was abnormally high. Mrs. C's illness was due to a lack of insulin, and is known as .... ——————————.

4. In Mrs. D's case, it was learned that she had had tuberculosis during her "teens." Now her resistance to infection was low, and she seemed unable to withstand stressful situations. The physician who was studying her case thought that she might be suffering from.............. ——————————.

VII. REFERENCES

Memmler, R. L., and Wood, D. L.: The Human Body in Health and Disease, ed. 4, pp. 243-253. Philadelphia, Lippincott, 1977.

Chaffee, E. E., and Greisheimer, E. M.: Basic Physiology and Anatomy, ed. 3, pp. 470-491. Philadelphia, Lippincott, 1974.

Ehrlich, P. R., Holm, R. W., and Soulé, M. E.: Introductory Biology, pp. 335-344. New York, McGraw-Hill, 1973.

Schmidt-Nielson, K.: Animal Physiology, pp. 642-651. New York, Cambridge University Press, 1975.

Stonehouse, B.: The Way Your Body Works, pp. 50-51. London, Beasley, 1974.

Villee, C. A., and Delthier, V. G.: Biological Principles and Processes, pp. 659-676. Philadelphia, Saunders, 1971.

CHAPTER NINETEEN

Reproduction

I. OVERVIEW

Through the functioning of the reproductive system, the continuation of the human race is assured. Human reproduction is *sexual* (as compared to the *asexual* reproduction of some of the simplest forms of life).

The male reproductive system consists of the *testes*, the *seminal vesicles*, the *prostate gland*, the *penis* and the *bulbourethral glands*; in the female this system comprises the *ovaries*, the *fallopian tubes*, the *uterus*, the *vagina*, the *vulva*, the *greater vestibular glands* and the *perineum*.

Normally, *fertilization* of a female *ovum* by a male *spermatozoon* results in *pregnancy*, the period of about 9 months during which an *embryo* forms and develops into a *fetus*. The ability of a woman to bear children ends with the menopause, the final cessation of menstruation.

II. TOPICS FOR REVIEW

1. sexual and asexual reproduction
2. characteristics common to male and female reproductive systems
3. male reproductive system
    a. testes
    b. tubes
    c. seminal vesicles
    d. prostate gland
    e. urethra and penis
    f. mucous glands
    g. spermatozoa
    h. disorders of male reproductive system
4. female reproductive system
    a. ovaries
    b. fallopian tubes
    c. uterus
    d. vagina
    e. vulvovaginal glands
    f. vulva and perineum
    g. disorders of female reproductive system
5. pregnancy
    a. first stages
    b. the embryo

        c. the fetus
        d. toxemias of pregnancy; discomforts associated with pregnancy
        e. parturition
        f. mammary glands; lactation
        g. multiple births
        h. disorders of pregnancy
        i. disturbances of lactation
    6. fetal deaths; live births
    7. menopause

III. MATCHING EXERCISES

Matching only within each group, print the answer in the space provided. An answer may be used more than once.

Group A

ovum	ovary	epididymis
gonads	asexual	testosterone
testis	scrotum	sexual
spermatozoa		

1. The specialized sex cells in the male are called........... _____.

2. Since the simplest forms of life require no partner in order to reproduce, they are said to be................. _____.

3. A characteristic shared by both men and women is the presence of sex glands, or............................ _____.

4. The specialized sex cell of the female is the.............. _____.

5. The female gonad is also known as the................. _____.

6. The male gonad is the.................................... _____.

7. The testes are normally located in a sac that is suspended between the thighs. This sac is the.................... _____.

8. The bulk of the tissue of the testes is arranged in tubules. Within the walls of these tubules there is the production of .... _____.

9. Groups of cells located between the tubules of the testes are responsible for the secretion of the male hormone.... _____.

10. The spermatozoa mature and become motile within a temporary storage area, a 20-foot tube, the.............. _____.

11. In most animals, there is differentiation into male and female; reproduction is therefore said to be............. _____.

Group B

seminal vesicles   ductus deferens   sex glands
ejaculatory ducts   gametes   urethra
spermatic cord   penis

1. The gonads are the .... _____.

2. Another term used in referring to the male and female sex cells is .... _____.

3. The straighter upward extension of the epididymis is the _____.

4. The combination of ductus deferens, the nerves and the blood and lymph vessels that extend from the scrotum and testes on each side is named the.................... _____.

5. Behind the urinary bladder in the male are 2 tortuous muscular tubes with glandular linings. These are the..... _____.

6. The vas deferens may also be called the................ _____.

7. The spermatozoa are carried in the seminal vesicle secretion through the prostate gland via 2 tubes called the.... _____.

8. A single tube conveys urine from the bladder and carries reproductive cells to the outside. This tube is the........ _____.

9. The external genitalia of the male include the scrotum and the .... _____.

10. Spermatozoa are nourished by secretions produced by the glandular lining of the............................... _____.

Group C

phimosis   erection   ductus (or vas) deferens
penis   Cowper's glands   orchitis
sterility   cryptorchidism   gonorrhea
prepuce   semen

1. In male ejaculation, a mixture of spermatozoa and secretions is expelled which is called..................... _____.

2. In the male, the longest part of the urethra extends through the spongelike .... _____.

225

3. The bulbourethral glands, which are pea-sized organs found in the pelvic floor tissues below the prostate gland of the male, are also known as.......................... _____.

4. Inflammation of the testes may be due to the organism of mumps or of tuberculosis. This disorder is called........., _____.

5. Normally, the testes descend into the scrotal sac during fetal life; should this not occur, the resulting condition would be known as................................ _____.

6. Burning and pain on urination, accompanied by a discharge from the urethra, are symptoms of the most common venereal disease in the male, namely........... _____.

7. In both the male and the female, infection that is not adequately treated may result in the inability to reproduce, or .... _____.

8. The foreskin is also called the......................... _____.

9. Tightness of the foreskin which prevents it from being drawn back is called................................ _____.

10. In order to prevent the spermatozoa from reaching the urethra, purposeful sterilization of the male is accomplished by removing a portion of the.................. _____.

11. Semen is expelled into the female vagina through the stiffening of the penis known as....................... _____.

Group D

vulva	vagina	fallopian tubes
ovaries	ovarian follicles	uterus
ovulation	vulvovaginal glands	fimbriae

1. Two structures, made of peritoneum and called the broad ligaments, serve as anchors for the.................... _____.

2. The sacs within which the ova mature are called the..... _____.

3. The rupture of an ovarian follicle permits an ovum to be discharged from the ovary surface. This is called........ _____.

4. The mature ovum travels from the region of the ovary into the oviducts or .... _____.

5. A current in the peritoneal fluid sweeps the ovum into the oviduct. This current is produced by the fringelike....... _____.

6. Before birth the fetus grows in a muscular organ located between the urinary bladder and the rectum. This organ is the .... _____.

7. Bartholin's glands, situated above and to each side of the vaginal opening, are also known as the greater vestibular glands and the .... _____.

8. Connecting the uterus with the outside is the lower part of the birth canal, the................................. _____.

9. The labia, the clitoris and related structures comprise the external parts of the female reproductive system which are called the .... _____.

Group E

carcinoma	cervix	amenorrhea
dysmenorrhea	fundus	endometrium
myomas	fornices	salpingitis
corpus		

1. Located above the level of the tubal entrances is the small rounded part of the uterus called the................. _____.

2. The upper part of the uterus is the larger part; it is called the body, or .... _____.

3. The necklike part of the uterus dips into the upper vagina; this necklike part is called the................. _____.

4. The specialized tissue that lines the uterus is known as... _____.

5. The cervix dips into the upper vagina so that a circular recess is formed; this gives rise to the spaces known as... _____.

6. Painful or difficult menstruation is called............... _____.

7. Absence of the menstrual flow is known as.............. _____.

8. About half of the women who reach the age of 50 have one or more uterine growths known as fibroids, or....... _____.

9. Inflammatory infection of the fallopian tubes is described as .... _____.

10. A test in which cells from the upper vagina are examined after being treated with Papanicolaou's stain is done to detect .... _____.

Group F

parturition	umbilicus	embryo
fetus	placenta	labor pains
afterbirth	toxemias of pregnancy	corpus luteum
amniotic sac	vernix caseosa	

1. The endometrium is prepared for the fertilized ovum by the hormone progesterone, which is produced by the..... _____.

2. Serving as the organ for nutrition, respiration and excretion for the embryo is a flat, circular structure called the _____.

3. Following fertilization of an ovum and until the end of the third month, the developing organism is called the.... _____.

4. From the end of the third month until birth the developing organism is known as the......................... _____.

5. The fetus is protected by a fluid produced by the membrane lining the .... _____.

6. Nature provides various protective mechanisms for the fetus. The cheesy material that protects the skin is known as .... _____.

7. The process of giving birth to a child is described by the term .... _____.

8. A small part of the umbilical cord remains attached to the navel for a few days following birth. The scientific name for the navel is the.............................. _____.

9. Normally, within half an hour after the child is born, the placenta together with the membranes of the amniotic sac and most of the umbilical cord are expelled as the.... _____.

10. Hypertension may be a sign of certain metabolic disorders that may develop during pregnancy. These disorders are classified as .... _____.

11. Parturition is usually divided into 3 stages. In the first stage the muscles of the uterus begin the contractions known as .... _____.

Group G

menstrual flow	posterior fornix	radical mastectomy
estrone	*Trichomonas vaginalis*	hysterectomy
menopause	external genitalia	leukorrhea

1. Removal of the uterus is known as.................... _____.

2. Congestion or infection of one or more parts of the female reproductive tract may cause a whitish vaginal discharge called .... _____.

3. Both ovarian hormones are involved in preparing the endometrium for pregnancy and both are carried by the bloodstream to the uterus. The preparation is initiated by the hormone found in the fluid surrounding the maturing ovum. This hormone is................................ _____.

4. The peritoneal cavity of the female is deepest behind the upper vaginal canal. This means that there is a thin wall separating the lower abdominal cavity from the upper vaginal canal. This dorsal space in the upper vagina is called the .... _____.

5. Bits of cast-off endometrium are found in the bloody discharge that is known as the............................ _____.

6. The vulva is also called the............................ _____.

7. A common causative organism of leukorrhea is a protozoan that is called................................. _____.

8. Cessation of ovarian activity brings about the period of life known as .... _____.

9. Cancer of the breast is the third most frequent malignant disorder in women; it is usually treated by the operation known as .... _____.

IV. LABELING

For each of the following illustrations, print the name or names of each labeled part on the numbered lines.

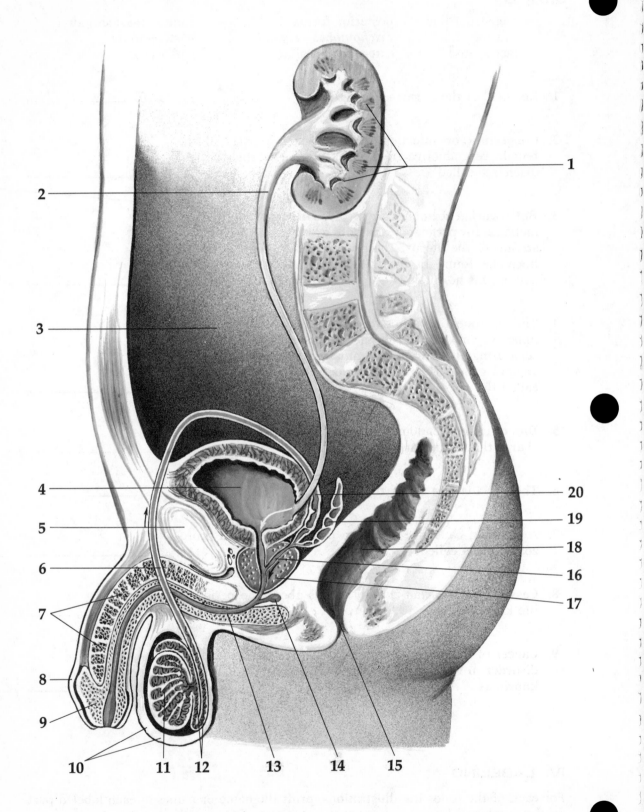

Male genitourinary system.

1. _____

2. _____

3. _____

4. _____

5. _____

6. _____

7. _____

8. _____

9. _____

10. _____

11. _____

12. _____

13. _____

14. _____

15. _____

16. _____

17. _____

18. _____

19. _____

20. _____

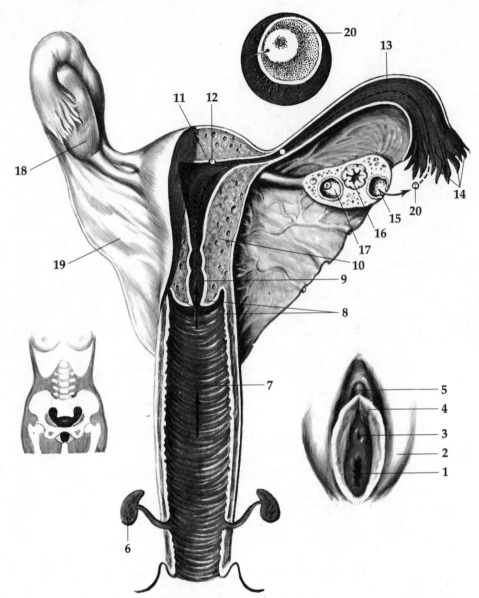

Female reproductive system.

1. _____   8. _____   14. _____

2. _____   9. _____   15. _____

3. _____   10. _____   16. _____

4. _____   11. _____   17. _____

5. _____   12. _____   18. _____

6. _____   13. _____   19. _____

7. _____   20. _____

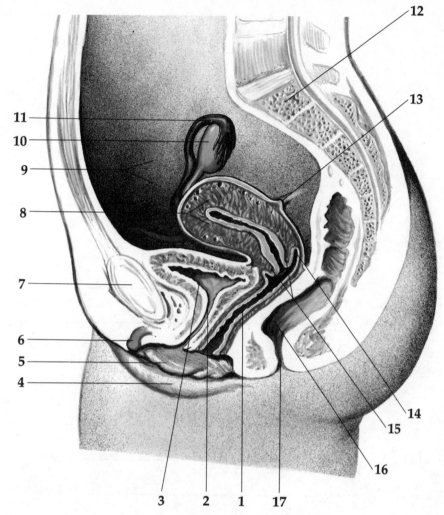

Female reproductive system, sagittal section.

1. _____	10. _____
2. _____	11. _____
3. _____	12. _____
4. _____	13. _____
5. _____	14. _____
6. _____	15. _____
7. _____	16. _____
8. _____	17. _____
9. _____	

1. _____
2. _____
3. _____
4. _____
5. _____
6. _____
7. _____
8. _____
9. _____
10. _____
11. _____
12. _____
13. _____
14. _____
15. _____
16. _____
17. _____

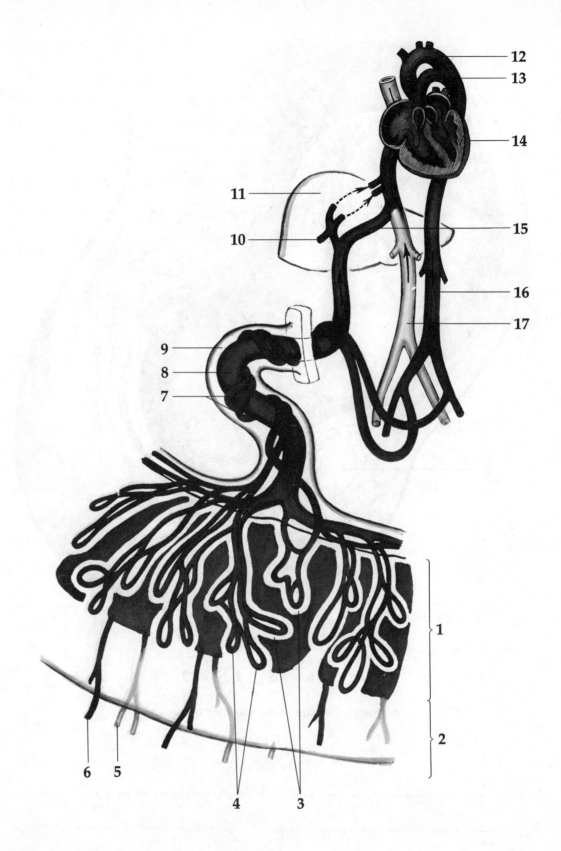

Fetal circulation and the placenta.

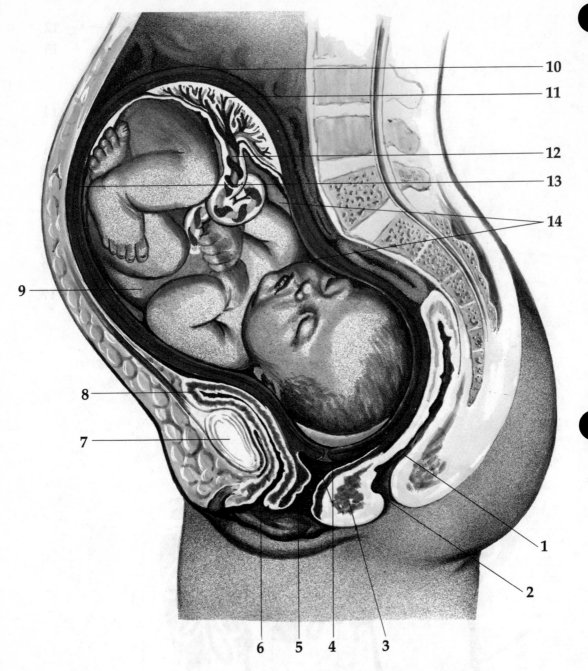

Midsaggital section of the pregnant uterus.

1. _____   4. _____

2. _____   5. _____

3. _____   6. _____

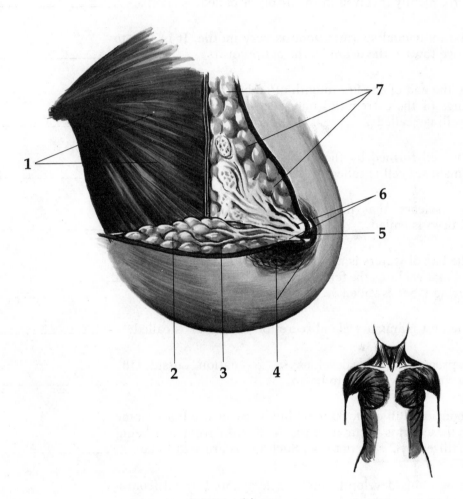

Section of breast.

1. _____
2. _____
3. _____
4. _____

5. _____
6. _____
7. _____

V. COMPLETION EXERCISE

Print the word or phrase that correctly completes the sentence.

1. The motility of the spermatozoa is maintained by several secretions including one from the seminal vesicles and one produced by the prostate gland called the .... _____ .

2. The region of the inguinal canal, being somewhat weak, is frequently involved in the disorder called .... _____ .

3. The individual spermatozoon is very motile. It is able to move toward the ovum by the action of its .... _____ .

4. By the end of the first month of embryonic life the beginnings of the extremities may be seen. These are 4 small swellings called .... _____ .

5. The cell formed by the union of a male sex cell and a female sex cell is called a .... _____ .

6. The science that deals with the development of the embryo is called .... _____ .

7. The bag of waters is a popular name for the membranous sac that encloses the fetus. The clear liquid that is released during labor is called .... _____ .

8. The first mammary gland secretion to appear is called ... _____ .

9. A pregnancy that develops in a location outside the uterine cavity is said to be an .... _____ .

10. About once in every 80 to 90 births twins are born. Some of these twins occur as a result of 2 different ova being fertilized by 2 spermatozoa. Such twins are said to be .... _____ .

11. Some twins develop from a single zygote formed from a single ovum that has been fertilized by a single spermatozoon. The embryonic cells separate during early stages of development. These twins are described as .... _____ .

12. The mammary glands of the female provide nourishment for the newborn through the secretion of milk; this is called .... _____ .

13. According to a classification approved by the World Health Assembly the loss of an embryo or fetus at any time during the first 20 weeks of pregnancy is termed a Group I or _____ .

238

14. To describe a newborn who breathes or shows any evidence of life such as a heart beat or pulsation of the umbilical cord, the classification used is................ _____.

15. In place of the older term, premature infant, the baby that weighs less than 5½ pounds (2500 grams) is called an _____.

16. The passage of the fetus through the cervical canal and the vagina to the outside takes place during labor's...... _____.

VI. PRACTICAL APPLICATIONS

Study each discussion. Then print the appropriate word or phrase in the space provided. The following patients were among those seen by a physician who specialized in women's diseases.

1. Mrs. G complained of soreness and discomfort of the breasts following the birth of her baby. The physician diagnosed her disorder as inflammation of the breast, or _____.

2. Mrs. K, age 43, thought that the bleeding she was now experiencing might be associated with early menopause. The physician examined her and found that a firm mass was present in the upper part of the uterus. This might be a myoma, which is commonly called a................ _____.

3. Miss C, age 16, said that she always felt irritable and depressed the week preceding her menses. A low salt diet and medication were prescribed to make her more comfortable during this period of........................ _____.

4. Because of episodes of hemorrhage during her pregnancy Mrs. M was hospitalized. There was a possibility that in Mrs. M's case the placenta was attached to the lower part of the uterus instead of the upper part. This condition is known as....................................... _____.

5. Because of the seriousness of Mrs. M's condition, an operation was considered; this would provide for delivery through an incision made in the abdominal wall and the wall of the uterus. This operation is called a............. _____.

VII. REFERENCES

Memmler, R. L., and Wood, D. L.: The Human Body in Health and Disease, ed. 4, pp. 255-272. Philadelphia, Lippincott, 1977.
Chaffee, E. E., and Greisheimer, E. M.: Basic Physiology and Anatomy, ed. 3, pp. 493-518. Philadelphia, Lippincott, 1974.
Ehrlich, P. R., Holm, R. W., and Soulé, M. E.: Introductory Biology, pp. 457-461. New York, McGraw-Hill, 1973.
Stonehouse, B.: The Way Your Body Works, pp. 54-63. London, Beasley, 1974.

CHAPTER TWENTY

Heredity and Hereditary Diseases

I. OVERVIEW

The scientific study of heredity is less than 2 centuries old. During the last few decades there have been brilliant and illuminating findings particularly related to the chromosomes and DNA; nevertheless many mysteries remain and there is much more to be studied. Gregor Mendel was the first person to put the study of heredity on a scientific footing. He called attention to the independent units of hereditary influence, which he called "factors" and which we now call "genes."

There are many thousands of genes in each cell nucleus. Each chromosome in the nucleus is composed of a complex molecule, DNA, and the genes are parts of this molecule. Genes direct enzyme action, which in turn makes possible the chemical reactions that comprise metabolism. Defective genes disrupt normal enzyme activity, with the result that congenital disorders, such as sickle cell anemia, albinism and phenylketonuria, can be manifest.

Genes are classified as dominant or recessive. If one parent contributes a dominant gene then all the offspring will show the trait, e.g., certain types of dwarfism (achondroplasia), extra fingers (polydactyly) and drooping eyelids (ptosis). Traits carried by recessive genes may remain hidden for generations and be revealed only if they are contributed by both parents, e.g., albinism, deaf-mutism (deafness from birth), microcephaly (abnormally small head) and sickle cell anemia. In some cases, treatment begun early may be helpful in preventing problems associated with the disorder. Genetic counseling should be sought by all potential parents whose relatives are known to have an inheritable disorder.

II. TOPICS FOR REVIEW

1. heredity
2. genes and chromosomes
3. DNA and RNA
4. pedigrees and karyotypes
5. genetic counseling
6. examples of genetic disorders

III. MATCHING EXERCISES

Matching only within each group, print the answer in the space provided. The same answer may be used more than once.

Group A

genes	congenital	DNA
chromosomes	life span	Mendel

1. Among the traits influenced by heredity is a person's..... _____.

2. A condition that exists from birth is said to be.......... _____.

3. Independent units of heredity are called............... _____.

4. The person who is credited with the first scientific investigation of heredity was an Austrian monk named........ _____.

5. A combination of thousands of genes is found in each of the nuclear structures known as...................... _____.

6. The complex molecule that comprises the chemical compound of each chromosome is........................ _____.

Group B

DNA	enzymes	replication
RNA	structural proteins	hereditary
mutation	X and Y	

1. During cell division (mitosis) each chromosome duplicates itself, a process called............................ _____.

2. Manufacture of structural proteins and of enzymes is controlled by the cellular chemical compound called..... _____.

3. Formation of the chemical compounds mentioned in question 2 is controlled by the genetic material called........ _____.

4. Traits transmitted via the genes are.................... _____.

5. The main components of skin and muscle are........... _____.

6. Proteins that promote chemical reactions within cells are _____.

7. Every normal cell of the male body contains 2 different sex chromosomes known as .... _____.

8. A spontaneous chromosomal change is called a......... _____.

Group C

phenylketonuria   karyotype   pedigree
lymphocytes   albinism   amniocentesis

1. The chromosomal elements typical of a cell form the..... _____.

2. A complete family history comprises a family tree or..... _____.

3. Removal of fluid from the sac surrounding the fetus is called .... _____.

4. An inherited disorder of metabolism in which an amino acid is involved is................................... _____.

5. Persons who lack skin and hair pigment are especially susceptible to skin cancer. They have inherited the condition _____.

6. Cells removed from the body, cultured (grown in a special medium) and then used for chromosome studies include .... _____.

IV. COMPLETION EXERCISE

Print the word or phrase that most accurately completes the sentence.

1. A spontaneous chromosomal alteration is called a(n)..... _____.

2. In Down's syndrome the cells have an extra............. _____.

3. In the disorder phenylketonuria there is a lack of a protein that governs metabolism of the amino acid phenylalanine. Such a protein functions as.................... _____.

4. The most frequent inherited metabolic disorder of the white race is .... _____.

5. Inherited muscular disorders, in which the chief feature is progressively severe muscle weakness, are known as... _____.

6. The inherited condition in which the skin and hair color is strikingly white is................................. _____.

V. PRACTICAL APPLICATIONS

Study each discussion. Then print the appropriate word or phrase in the space provided.

Group A

These patients were seen in a pediatric clinic.

1. A rigid diet had been prescribed for infant M after it had been determined that he had an inherited disorder characterized by a lack of the enzyme required for metabolism of phenylalanine. Infant M's disorder is called.......... _____.

2. A black child about 4 years of age was brought to the hospital with a history of swelling and pain in the joints of his hands and feet. Among the studies done was one that revealed crescent-shaped red blood cells. It was determined that his condition was a hereditary one called..... _____.

3. A white child, age 6, had a history of difficulty in breathing, a digestive disorder for which a pancreatic extract was prescribed, and frequent respiratory infections. These problems are typical of a hereditary disorder found mostly in caucasians, namely .... _____.

4. Baby D's face was round; her eyes were close-set and slanted upward at the sides. Her appearance was oriental. The infant's condition, commonly called mongolism, is correctly termed .... _____.

5. Young Mrs. D brought her 2-year-old son in so that his response to the prescribed diet and administration of the vitamin thiamin could be studied. In his hereditary disorder, the urine has such a peculiar odor that the disease is often called .... _____.

Group B

In a University clinic studies of several hereditary disorders were under way. Read each discussion and respond in the space provided.

1. Mrs. Y, age 34, was pregnant a second time. Her first child had Down's syndrome, and she was apprehensive lest a second child would be similarly affected. A study of the fluid in the sac surrounding the embryo was ordered. This procedure is called a(n)........................ _____.

2. The cells from Mrs. Y's amniotic fluid were grown (cultured) and then stained for special study. A diagnostic indication of Down's syndrome was made, namely the presence of an extra .... _____.

3. Mrs. G and her husband received genetic counseling. They had one normal child and one albino. Was there a possibility that a third child would be an albino also? Neither parent was an albino and they knew of no relatives who were. A trait carried for generations without being manifest is said to be.......................... _____.

4. Mr. H, age 25, was scheduled for a check-up and discussion of his diet and medication. He had a hereditary disorder, Wilson's disease, in which there is abnormal accumulation of the metallic ion...................... _____.

5. Mr. H was warned by his physician that continued disregard of his dietary prescription and prescribed medication would lead to damage to the largest organ of the abdomen, namely the .... _____.

6. Baby K's recent examination indicated good progress because his mother was following the physician's instructions about diet. Baby K had a deficiency of the enzyme phenylalanine, one of the protein building blocks known as .... _____.

7. Baby G was brought in for further study and an evaluation of his medications. Along with strict dietary control he was receiving vitamin B or........................ _____.

VI. REFERENCES

Memmler, R. L., and Wood, D. L.: The Human Body in Health and Disease, ed. 4, pp. 273-278. Philadelphia, Lippincott, 1977.
Ehrlich, P. R., Holm, R. W., and Soulé, M. E.: Introductory Biology, pp. 83-95, 102-123. New York, McGraw-Hill, 1973.
Frobisher, M.: Fundamentals of Microbiology, ed. 9, pp. 252-254. Philadelphia, Saunders, 1974.
Gardner, E. J.: Principles of Genetics, ed. 5. New York, Wiley, 1975.
Jolly, E.: The Invisible Chain: Diseases Passed on by Inheritance, pp. 1-214. Chicago, Nelson-Hall, 1972.
Nason, A., and Dehaan, R. L.: The Biological World, pp. 8-9, 102-103, 189-235. New York, Wiley, 1973.

CHAPTER TWENTY-ONE

Immunity, Vaccines and Serums

I. OVERVIEW

Although the body is constantly being exposed to pathogenic organisms, infection develops relatively rarely. This is because the body has many "lines of defense" against pathogenic invasion. The intact *skin* and *mucous membranes* serve as mechanical barriers, as do certain *reflexes* such as sneezing and coughing. Next is the process of *inflammation*, by which the body tries to get rid of the irritant or to minimize its harmful effects. Finally, the ultimate defense is *immunity*. There are two kinds of *immunity*. *Inborn* immunity is inherited; it may exist on the basis of *species, racial,* or *individual* characteristics. *Acquired* immunity may be of the *natural* type (i.e., acquired before birth, or by contacting the disease); or of the *artificial* type (i.e., provided by a *vaccine* or an *immune serum*). The chemical processes in *allergy* are similar to those of immunity. The *rejection syndrome* that often takes place following organ *transplantation* might be compared to a greatly increased allergic reaction.

II. TOPICS FOR REVIEW

1. basis of infection
2. body's defenses against infection
3. immunity
   a. inborn
   b. acquired
4. vaccines and serums
5. allergy
6. transplantation; rejection syndrome

III. MATCHING EXERCISES

Matching only within each group, print the answer in the space provided. An answer may be used more than once.

Group A

nerve tissue	virulence	nonspecific resistance
immunity	infection	portal of entry
toxins	respiratory tract	

1. The means by which a pathogenic organism invades the body is called the.................................. _____.

2. The degree to which an organism can cause disease and its power to overcome body defenses are known as its... _____.

3. The unbroken skin and mucous membranes are elements of the body's defenses against any harmful agent. Such protective devices are part of the body's.............. _____.

4. Some pathogens have an affinity, or preference, for certain tissues. The virus of poliomyelitis, for example, attacks only .... _____.

5. A second example of tissue preference is seen in the common cold and influenza viruses, which invade the........ _____.

6. A break in the skin through which pathogenic organisms can easily reach deeper tissues usually results in......... _____.

7. The poisons produced by pathogens are known as....... _____.

8. The body's ability to defend itself against a certain specific agent is spoken of as its specific resistance, or....... _____.

Group B

inborn immunity	inflammatory reaction	antigen
species immunity	racial immunity	attenuation
active immunity	passive immunity	

1. The body tries to get rid of (or minimize the effects of) an irritant by a series of defensive responses that constitute an .... _____.

2. Animals are susceptible to diseases that do not affect man; the reverse is also true. In other words, both possess a... _____.

3. The greater resistance of black Americans to malaria and yellow fever over white Americans is an example of...... _____.

4. Any foreign substance introduced into the body by a pathogen provokes a response. The foreign substance is known as an .... _____.

5. An inherited immunity is usually called a(n)............ _____.

6. A person who is infected by a pathogen or its toxin produces antibodies that make him immune to that infection for a long period of time, perhaps for life. This type of immunity is called .... _____.

248

7. In the type of immunity provided by vaccination, a weakened pathogen or toxin is injected into the body. The process of reducing the virulence of the pathogen or toxin is called .... _____.

8. The resistance of the newborn infant to contagious diseases is due to a type of immunity classified as........... _____.

Group C

inoculation	artificially acquired	antigens
serum	serum sickness	weakened
allergy	vaccination	gamma globulin

1. A person who has been bitten by an animal suspected of having rabies is given the Pasteur treatment, a form of... _____.

2. Some Rh negative persons may become sensitized to the Rh protein. In such cases the Rh factor serves as an example of substances known as........................ _____.

3. The administration of a vaccine results in the type of immunity classified as .... _____.

4. Vaccines contain pathogens or toxins that have been attenuated or .... _____.

5. The mixture of antibodies and blood plasma that contains no fibrinogen is known as............................ _____.

6. A substance that is of human origin, rather than animal origin, is rich in antibodies against measles. This substance is called .... _____.

7. A reaction between an antigen and an antibody that takes place within the cells of sensitized tissue results in....... _____.

8. The general term that refers to the production of allergy by a serum is .... _____.

9. Another word that has the same meaning as vaccination is _____.

Group D

convalescent serum	antivenin	antitoxin
toxins	allergens	rejection syndrome
sensitized	antibodies	

1. Protection against measles is provided by gamma globulin through its .... _____.

2. The reason horses are widely used in the production of immune serums is that their tissues produce large quantities of antibodies when injected with pathogens or their . . _____.

3. Diphtheria serum is an example of an immune serum. It contains large amounts of the type of antibody called an _____.

4. The type of serum that is injected to combat the effects of bites of various poisonous snakes is called . . . . _____.

5. In susceptible persons, repeated exposure to an allergen will cause his tissues to become . . . . _____.

6. Pollens and house dust contain proteins which in susceptible persons act as . . . . _____.

7. Hay fever is an example of disorders due to . . . . _____.

8. The serum of a person who is recovering from a certain disease may be administered to another person to provide passive immunity. Serum so used is known as . . . . _____.

9. The natural tendency for every organism to reject foreign substances has given rise to a serious problem in organ transplantation. This obstruction which frequently prevents a successful graft is referred to as the . . . . _____.

IV. COMPLETION EXERCISE

Print the word or phrase that correctly completes the sentence.

1. The mixture of leukocytes and fluid from the blood plasma which is produced as the body tries to defend itself against pathogens is known as the . . . . _____.

2. One aspect of a pathogen's virulence has to do with its ability to overwhelm the host; this may be called its aggressiveness, or . . . . _____.

3. The sum of the body's natural defenses against disease is known as its . . . . _____.

4. The first line of defense against disease includes mechanical methods for removing foreign matter from the respiratory tract. Among these are sneezing and coughing, which are . . . . _____.

5. The action of leukocytes in which they engulf and digest invading pathogens is known as . . . . _____.

6. Living and dead leukocytes and pathogens, exudate and destroyed tissue form a yellowish, thick fluid called...... _____.

7. Immunity is a selective process through which a person may be immune to one disease but not to another. This selective characteristic is called .... _____.

8. The virulence of an invading organism is partly dependent on its ability to produce toxins, or..................... _____.

9. The word that means the opposite of immunity is........ _____.

10. There are 2 main categories of immunity. One is inborn immunity while the other is......................... _____.

11. Heat, redness, swelling and pain are considered the classic symptoms of the .... _____.

12. Some persons have greater resistance to disease than the other members of their group; such persons are said to have .... _____.

13. An artificially acquired immunity may be provided for a limited time by injecting an immune serum or for longer by using a .... _____.

14. Antibodies transmitted from the mother's blood to the fetus provide a type of short-term borrowed immunity called .... _____.

15. The administration of vaccine, on the other hand, stimulates the body to produce a longer lasting type of immunity called ..../................. _____.

16. The irritating effects of the allergic response is believed to be an antigen-antibody reaction that results in the liberation of a substance called a..................... _____.

17. Although German measles is a very mild disease, children should be vaccinated in order to prevent the possibility of a pregnant woman contracting it. The use of rubella vaccine will lower the number of..................... _____.

V. PRACTICAL APPLICATIONS

Study each discussion. Then print the appropriate word or phrase in the space provided.

1. Mr. O brought his 38-year-old wife to the office for emergency treatment following a bee sting. Since Mrs. O was in a state of near-shock, the physician administered an injection of epinephrine; Mrs. O's condition improved. The physician then made several suggestions to help Mrs. O deal with her sensitivity to bee stings which he described as a(n) .... _____.

2. Eight-year-old K complained of swelling and pain in the area of a wound he received when he stepped on a nail several days earlier. Because of the possibility of tetanus, K was given an injection to neutralize the tetanus toxin. This was an injection of............................. _____.

3. Mrs. R brought her 2-month-old infant to the office for the first of a series of injections to inoculate him against several serious diseases. The vaccine used for this purpose contains the weakened toxins of the organisms causing these diseases. This type of vaccine is known as...... _____.

4. Miss Y was allergic to pollens. In the hope that her tissues would be desensitized the physician was giving her repeated injections of the............................. _____.

VI. REFERENCES

Memmler, R. L., and Wood, D. L.: The Human Body in Health and Disease, ed. 4, pp. 279-287. Philadelphia, Lippincott, 1977.

Frobisher, M.: Fundamentals of Microbiology, ed. 9, pp. 224-226, 392-407. Philadelphia, Saunders, 1974.

Glasser, R. J.: The Body Is the Hero. In Reader's Digest, October, 1976, pp. 252-290.

Nason, A., and Dehaan, R. L.: The Biological World, pp. 408-410. New York, Wiley, 1973.

Salle, A.: Fundamental Principles of Bacteriology, ed. 7, pp. 851-881. Los Angeles, McGraw-Hill, 1973.

Stonehouse, B.: The Way Your Body Works, pp. 76-77. London, Beasley, 1974.